The Merino and Anglo-Merino Breeds of Sheep

by Charles Henry Hunt

with an introduction by Jackson Chambers

This work contains material that was originally published in 1809.

This publication is within the Public Domain.

Self Reliance Books

Get more historic titles on animal and stock breeding, gardening and old fashioned skills by visiting us at:

http://selfreliancebooks.blogspot.com/

Introduction

I am pleased to present yet another practical title on breeding and raising livestock.

The work is in the Public Domain and is re-printed here in accordance with Federal Laws.

As with all reprinted books of this age that are intended to perfectly reproduce the original edition, considerable pains and effort had to be undertaken to correct fading and sometimes outright damage to existing proofs of this title. At times, this task is quite monumental, requiring an almost total "rebuilding" of some pages from digital proofs of multiple copies. Despite this, imperfections still sometimes exist in the final proof and may detract from the visual appearance of the text.

I hope you enjoy reading this book as much as I enjoyed making it available to readers again.

Jackson Chambers

TO

THE REV. JAMES WILLIS,

VICAR OF SOPLEY,

FOUNDER AND PRESIDENT OF THE CHRISTCHURCH AGRICULTURAL SOCIETY,

AND

ONE OF HIS MAJESTY'S JUSTICES OF THE PEACE FOR THE COUNTY OF HANTS.

IN the very distinguished situation which you have so long held as President of the Christchurch Agricultural Society, with equal credit to yourself, and advantage to the neighbourhood, and to the public, having, as a member of that most respectable Society, for several years witnessed your zeal and patriotic ardour in all concerns connected with Agriculture and the public weal, and particularly in the Merino cause; and as several gentlemen of high consideration in this county, and most of them members of *your* Society, have, as well as myself, become considerable Merino proprietors, I feel I cannot inscribe my humble labours in the Merino fold to any one with more propriety than yourself.

Being

Being conscious that the following Treatise, which is published by me with no other motives than to afford information to the practical farmer, and the hope of promoting the public good, I trust it will meet with an indulgent reception; which, from the hasty manner in which it has been written, I am satisfied it will require even from my friends. Happy in this opportunity of acknowledging the very friendly and flattering attention you have on all occasions shewn me, I beg leave to add, it will be highly gratifying to me to find the following pages meet with your approbation and countenance; and remain with great regard,

My dear Sir,

Your very sincere friend and very

faithful humble servant,

THE AUTHOR.

LIST OF MERINO AND ANGLO-MERINO PROPRIETORS,

IN THE COUNTY OF HANTS.

* SIR GEORGE TAPPS, BART. Hinton Admiral, near Christchurch.

* GEORGE H. ROSE, ESQ. M.P. Mudeford, near Christchurch.

* CHARLES JENKINSON, ESQ. M.P. Beech House, near Christchurch.

* GENERAL CAMERON, Belvidere House, near Christchurch.

* COLONEL CUNYNGHAME, Maltshanger, near Basingstoke.

WM. MITFORD, ESQ. Exbury House, near Beaulieu.

COLONEL SEELE, Chilworth Lodge, near Southampton.

* The REV. Dr. WYNDHAM, Hinton, near Christchurch.

* The REV. JAMES WILLIS, Sopley, near Christchurch.

GEORGE EYRE, ESQ. Warrens, near Lyndhurst.

CHARLES MITCHELL, ESQ. Netherwood, do.

* CHARLES H. HUNT, ESQ. Belvidere Cottage, near Christchurch.

* CORNELIUS TRIM, ESQ. Throop, near Christchurch.

* GEORGE COMPTON, ESQ. Chilworth, near Southampton.

The REV. PETER DEBARY, Eversley, near Harford Bridge.

The REV. J. ESSEN, Exton, near Alton.

Those marked thus * are Members of the Christchurch Agricultural Society.

CONTENTS.

CONTENTS.

ON

ON THE

MERINO BREED

OF

SHEEP.

INTRODUCTORY CHAPTER.

THE practicability of producing in the British
isles, fine clothing wool, equal in quality to that
imported from Spain, has already been sufficiently demonstrated, in the publications of several
eminent characters whose attention has very pa-
triotically been devoted to the subject; and necessity has at length awakened us to a just sense
of its importance. Its importance, in fact, has
been most strikingly exemplified by the enormous
advances which have lately taken place in the
price of Spanish wool; and we are now feelingly
alive to the impolicy of depending on foreign

B   countries

countries for a commodity, of which an abundant supply for the purposes of our manufactures, may in process of time be obtained from the cultivation of our national resources.

The facts and observations which have from time to time been communicated to the public, by the highly respectable individuals* to whom I have just alluded, have at length overcome the prejudices which till lately had retarded in this country, the dissemination of the Spanish, or, as it is usually called, the Merino breed of sheep; and the improvement of our native breeds, by an intermixture with the Spanish, in the essential qualities

---

* Amongst whom may be more particularly enumerated the Right Honourable Lord Somerville, the Right Honourable Sir Joseph Banks, Dr. Parry of Bath, George Tollet, Esq. of Staffordshire, Edward Sheppard, Esq. of Uley in Gloucestershire, and other gentlemen of eminence, who have treated the subject with a degree of ability commensurate with its importance. Much interesting information has also been communicated on this subject by M. Lasteyrie, an ingenious French author, in his " History of the Introduction of Merino Sheep into the different States of Europe," M. Pictet, of Geneva, and other eminent foreigners.

qualities of an article on which our principal staple manufacture depends.

But as it is not often in the way of the practical farmer to meet with the several publications through which this important information is diffused, it has occurred to me, that his attention would be more easily gained to the subject by a short treatise, in which are concentred the most material facts, and such as have been corroborated by long experience and tolerably extensive practice.

Under this impression, and agreeing with Dr. Parry, that ' it is the duty of every one to use his utmost efforts to promote the interests of mankind, without fear of obloquy, or any other personal inconvenience,' my immediate object in the following pages, is to state briefly such useful and important facts as I have collected, and are confirmed by my own experience, and in the most plain language to place them in such a point of view, as may convince the practical farmer of their truth; or if he has still any doubts, to furnish him with

the

the ready means of removing them by personal inquiry and ocular demonstration. And to prove (which I trust I shall do satisfactorily to all those who are open to conviction) that Merino sheep, and their different crosses with our native breeds, do succeed in this country beyond the expectations of the most sanguine of the original breeders. That those imported here do not degenerate in any respect; that their wool, after being here many years, is as fine as it was the day they reached our shores; that the wool of their offspring produced here, is as fine as that of any of their progenitors; that all those of the pure blood, bred in England, as well as all the different crosses, are not only as hardy, or more so, and equally, or more healthy, and long lived, than any breed now in general use; and that they are also full as profitable to the farmer and grazier, as any such English breeds, in point of carcase, disposition to fatten, folding, &c. and, in short, in every other advantage to be derived from sheep; and that their mutton is not to be excelled.

Though all the objections started to the Merinos,

rinos, on their introduction by the paternal care of his Majesty above 16 years since, are now entirely removed, except as to their shape, which still gives great offence to the Leicestershire and South Down breeders, or as they pretend, affords them great amusement; I shall (as I write chiefly for the information of the inexperienced on this subject) review the several objections as they arise, in my progress through the following pages.

Having occasion to mention his Majesty's name, I cannot proceed farther without declaring, that though undoubtedly no King who ever reigned is more entitled to the blessings of his subjects, for his uniform and steady attention to, and for the many sacrifices he has made for, their interest and welfare, and for the examples he has set during a long and arduous reign, as a monarch, a father, and a man, whose numerous virtues are deeply impressed in the hearts of his subjects, and will for ages survive him, yet I am persuaded the day will come, when the introduction of Merino sheep, and the care and anxiety he has manifested to disseminate them through the kingdom, will

not

not be considered as the least beneficial act of his reign. His Majesty's plan of permitting a certain number of his pure Merino ewes and rams to be sold annually, by auction, by which the meanest of his subjects, without any interest or introduction, may become possessed of any he pleases, is highly beneficial.

This auction, from its institution, has been conducted upon principles so liberal, by the Right Hon. Sir Joseph Banks, who generally attended in person, and since he has from ill health declined the management of the Merino flock at Kew, (originally under his care) by Mr. Snart, his Majesty's Bailiff, that no person I believe ever returned from it dissatisfied.

It has been suggested by some, possibly those beginning to smart under the prevailing increase of Merino and Anglo-Merino flocks, that an auction is beneath the royal dignity; let such cavillers propose a better mode of carrying his Majesty's benign intentions into effect, which no doubt would be attended to, or rather let him visit these auctions, where he will find that

nothing

nothing can exceed the propriety with which
the whole business is conducted, or the urba-
nity with which all persons are attended to,
and every information and accommodation afforded, by Mr. Snart, and those under his direc-
tions. No narrow policy, or contrivance to
enhance the price, so usual at common auctions
and sales, is adopted here, nor are the sheep
produced in a pampered state, or trimmed up so
as to appear to the best advantage; on the contrary, they are as they are taken from the flock
a day or two before, and generally in low condition, and so far from using means to sell them
at a high price, I have more than once heard
both Sir Joseph Banks and others declare their
regret that the eagerness of the public to procure them, has lately run up the sheep to such
very high prices, least it should tend to defeat
the object of his Majesty to promote their circulation in thus offering them to the public,

The following is the average of prices which
these animals produced at the last three sales :—

	£. s. d.		£. s. d.
In 1806 Ewes averaged	12 7 0	Rams	14 17 0
1807	20 2 0	—	24 15 6
1808	23 12 5	—	33 10 1

Of

Of these, the rams had good mouths; the ewes were full-mouthed, and several broken. The highest price given for a ram at the last sale. was £74 : 11s. and for an Ewe, £38 : 17s. The lowest price given for a ram, was 18 guineas, and for an ewe 20 guineas. Seventeen Rams produced £571 : 11s. 6d. and twenty-five Ewes, £620 : 19s. together, £1192 : 10s. 6d.

Next to his Majesty, Lord Somerville has, I believe, the largest flock of pure Merinos, but his Lordship's are not precisely of the same breed as his Majesty's. This patriotic nobleman went himself to Spain several years since, and surmounting every obstacle, procured the best woolled ewes and rams which that country afforded, from the several different flocks. He has also, by letting his valuable rams to hire, in different parts of England and Scotland, contributed very largely to the distribution of this race, and to their good qualities being more generally known; and he has also, by his attention and skill in breeding, in several specimens I have lately seen of his rams and flock, very much indeed improved their shape.

Dr.

Dr. Parry has a very valuable Anglo-Merino flock; which, as will be mentioned hereafter, and may be more fully seen by referring to his several works, is so much improved by repeated crossings, and breeding in and in, as in many instances to have produced finer wool than that of any of the pure blood imported from Spain; and his sheep are also so much improved in shape, as to induce him to decline any further use of the pure Merino ram, which he is persuaded could not improve his wool, and might injure the shape he has acquired. In his excellent treatise on this subject, published in 1800, and entitled " Facts and Observations on the practicability and advantage of producing in the British Isles, Clothing Wool equal to that of Spain," he not only clearly demonstrates that he has himself done so, to the entire satisfaction and conviction of the Bath and West of England Agricultural Society, but in that and his subsequent publications, has so completely answered and removed the objections of the manufacturers to English grown wool, that they now all purchase our Merino, and Anglo-Merino fleeces, of every cross, with avidity, at astonishing prices.

It

It is natural that those persons who are deeply interested in other breeds should use all the means in their power to prevent the further increase of Merino and Anglo-Merino sheep. They consequently eagerly receive, and industriously circulate every tale they can pick up to their disadvantage. To obviate such effects, I shall only recommend to those persons who may find any thing in the following pages relating to these sheep, which strikes them as worthy their attention, to enquire for, and take the trouble of examining, the nearest Merino flocks, a satisfaction which I have no doubt every gentleman possessing one will readily afford them; and I am equally persuaded they will find not only a complete refutation of all the calumnies raised, but a confirmation of all I assert. The names of many noblemen and gentlemen possessing them will be found in the course of this work; and so great is the conviction of their excellence in the county of Hants, that not only many gentlemen, as will be seen at the beginning of this book, possess them, but many of the principal South-Down flockmasters have crossed very largely with the Merino rams. I have this year had applications from gentlemen farmers

of

of this description for more rams than I could myself supply, though I had many more than I had any idea I could dispose of or use; and I have reason to know that such applications have not been confined to me.

As I know some persons are deterred from adopting the Merino breed, from an idea that the market will be overstocked, it will not be amiss to observe here, that it appears from Parliamentary documents quoted by Dr. Parry, that in three years, ending in 1804, there were imported into this country 18,467,718 lbs. of Spanish wool, of which 16,986,644 lbs. were received directly from Spain; and that consequently, during these three years, Great Britain paid to foreign countries for the wool which was the basis of its fine woollen manufactures, at least £.4,700,000 or upwards of £.1,560,000 per annum.* To supply the whole quantity purchased at these prices, would, I may venture to assert, require upwards of a million and a half of sheep.

Could

---

* Dr. Parry's Essay on Merino Sheep, communicated to the Board of Agriculture. Vide their Communications. *Vol. 5, p.* 341.

Could any thing further be necessary to quiet the fears of those afraid of the markets being overstocked with fine wool, a circumstance from the foregoing statement not very likely to happen in the time of any person now alive, the following facts will, I trust, entirely have that effect. A question was proposed by a very eminent Gloucestershire clothier, to a member of the Bath Society who stood single in his avowed opposition to the amelioration of clothing wool, to the following effect:— " Did he know *any single article* manufactured from *coarse*, which would not *be better made from finer wool, if more was grown,* so *that it might be had equally cheap?"** No answer was attempted, or that I have found has since been given. And in a letter of a reverend gentleman, also mentioned by Lord Somerville, is the following observation:—" If it should be argued, that in time the fine woolled sheep might supersede others, it is some comfort to know that there is *not a single manufacture* in which *coarse wool is used,* that *would not be doubly valuable by being composed of that which is fine."*

Since

---

* Lord Somerville's Facts and Observations on Sheep, Wool, &c. published in 1803, p. 61.

Since the late very high and unreasonable price required, and in many instances given for Merino wool, from its scarcity, and the apprehension of its becoming much more so, I have very frequently observed the noblemen, gentlemen, and landlords, called upon in news-papers, magazines, &c. to exert themselves, and endeavour to enforce as far as they can by example, the patriotic measure of wearing none but woollen cloths manufactured from the fleeces of British sheep, which it is suggested would have a great effect upon trade, and tend to lower the exorbitant prices exacted for fine cloths. To this cry I very sincerely add my voice, but instead of incommoding themselves with coarse, heavy, cumbrous cloths, hardly to be endured in warm seasons, let them encourage, by every means in their power, the Merino and Anglo-Merino sheep, and we shall much sooner obtain the object sought after, of reducing the price of fine wools to their old standard, and derive all the advantages proposed without any of the inconveniencies.

CHAP.

CHAP. II.

Containing a description of Merino Sheep, with an account of their management in Spain; and shewing that the Spanish mode of treatment is by no means necessary in this country; and that in Spain it arises more from necessity and ancient custom than from any advantage to be derived from it, in the improvement of their wool; and also, that this climate is in every respect as well suited to them as that of Spain.

FOR the meaning of the word Merino, and whether the breed of Sheep so called were originally produced in Spain, or came from Britain, (as has been asserted) or from Italy, or any other country, it is so completely immaterial to the practical English farmer, for whose information I propose principally to write, that I must beg leave to refer persons inclined to satisfy themselves on such points, and to search deeper into Merino history, to the several authors

I have

I have alluded to in a preceeding page, who have written upon this subject; but more particularly to the able works of Dr. Parry, of Bath, which are replete with every source of information that can be required, and to whose patriotic exertions his country is most peculiarly indebted.

The Merino Sheep, in Spain, are about the size of our old South Down, and rather larger than the pure Ryeland. Their shape, though what the greatest painters have chosen as models,* is certainly not such as the English sheep-fanciers of the present day can admire. They are in general rather high on their legs, flat sided, and narrow across the loins, and consequently defective in the hinder quarter: it may however be observed that they are not all so, and there are many what we might call hand-some individuals in most flocks, of which, by a judicious selection and breeding in and in, there is no doubt but a good shape would soon be obtained, though it must be admitted that

crossing

* Dr. Parry. Communications to the Board of Agricul-, ture, *Vol.* v. *p.* 342.

crossing with our own good-shaped sheep is by far the readiest mode.

This defect of shape is, however, much counterbalanced by the peculiar quality of their skins, which will certainly please the grazier, as they are remarkably soft and loose, affording that evidence of a strong disposition to fatten, called PROOF. The skin is also of a very fair and flesh-coloured hue, which is particularly visible on the eyelids and lips, the only parts free from wool, in which these sheep are in general almost enveloped. It grows on the forehead as low as the eyes, and on their cheeks, covers their bellies, grows down to their hoofs on their hind legs, and sometimes on their fore legs, but this is not always the case. I have a ram, purchased at his Majesty's Sale in 1807, for which I have refused a hundred guineas, who, though he has usually thrown off upwards of 7lbs. of washed wool, which is remarkably fine, has none upon his legs: but his stock, particularly while hogs (or tegs) though shorn as lambs, are all well woolled on every part, and so much so about the cheeks and eyes, that I have been obliged to cut the wool from several of

them,

them, to enable them to see properly. According to M. Lasteyrie, one Merino sheep with another in France, gives about 5¼lbs. of wool, English weight, and this is supposed by Dr. Parry to be the average weight of ewes' fleeces in Spain. The length of the staple is from two to three inches, and differing from all other sheep, is as fine on the rump as the shoulder, or very nearly so. The wool of the ram is generally considered the coarsest and longest; that of the ewe, finest and shortest; and that of the wether, in both respects, between the two former. The average weight of rams' fleeces in Spain is estimated at about 7lbs. but this must, in general, in great measure depend upon the size of the ram. In England, when the sheep is in good condition, it is oftener more than less. Mr. Tollet has published, in the Annals of Agriculture, an account of his Merino flock, by which it appears that his pure blood in 1804, consisted of 16 rams, 32 ewes, and 8 shearling ewes; in all 56. The rams yielded of wool, in the yolk, 130lb. or 8lb. 2oz. each; the ewes, 191¾lb. or 5lb. 15oz. each; and the shear-

c   lings,

lings, 34¼ lb. or 4 lb. 5½ oz. each; exclusively of the lamb's fleece. The total weight was 356¼ lb. and the average 6 lb. 6 oz. each fleece. The 356¼ lb. were reduced, by washing to the Spanish state, to 184 lb.; and if, when purified by clean scowering, this wool, like the best Spanish, suffered a farther waste of 3½ lb. per score, the total reduction would be to 152 lb. or to somewhat less than 4-9ths of the original weight. It sold for 20s. per fleece, or 3s. 1½d. per lb. through all the sorts in the yolk; the value being taken from Spanish refina at only 6s. 4d. per lb.[*]

There seems to be a difference of opinion in some of the writers on this subject, as to which of the Spanish flocks have the finest wool. Sir Joseph Banks, on whose authority I am most disposed to rely, gives the palm to the PAULAR; and next to that the NEGRETE, from which all the rams and sheep which his Majesty had, and has disposed of since 1791, are descended; and these, both Lasteyrie and Bourgoing agree in describing as of superior size. Sir Joseph, on

---

[*] Dr. Parry. Communications to the Board of Agriculture, vol. v. p. 431.

on this occasion, quotes these two authors, the former of whom lived many years in Spain, and paid great attention to Merino sheep. They both agree too in asserting, that the three piles of PAULAR, NEGRETE, and ESCURIAL, have from their excellence been withheld from exportation, and retained for the royal manufactory at Guadalaxara ever since it was first established.

The rams are in general, horned; and the ewes, polled, or without horns. But it is not at all uncommon to meet with varieties in this respect: I purchased a ram without horns, or the smallest appearance of any, at the King's last sale, at a high price, and I have another of the same description, bred by Mr. Fane, of Oxfordshire: and I observed several ewes of the last importation, with small horns. The horns of many of his Majesty's rams are very light, and some very short and blunt, which are called snags; and I fancy these are descended from a very famous woolled ram, well known at Kew by the name of old Snags.

The

The number of Merino sheep in Spain, is estimated at about 6,000,000. These comprise a great number of flocks, belonging to different proprietors, who are chiefly grandees, or societies of monks. Many of these flocks differing in form, size, and fineness of wool, appear to be distinct varieties of the same race. A considerable part of Estremadura, Leon, and the neighbouring provinces of Spain, is appropriated to their maintenance, as are also broad green roads, seventy-five yards wide, leading from one province to another, with extensive resting places, where the sheep are baited on the road. And of so much importance is their welfare considered, and so extremely strict is the police of the country to preserve them during their journeys, from all hazard of disturbance or interruption, that no person, not even a foot passenger, who does not belong to the flocks, is suffered to travel upon these roads while the sheep are in motion.*

As soon as the pasturage, from the increasing heats

---

* Sir Joseph Banks's Letter to Sir John Sinclair, President of the Board of Agriculture, 18 Feb. 1809.

heats of April and May, becomes scanty, the sheep begin their march towards the mountains of Leon, and after having been shorn on the road, at vast establishments called Esquileos, (where they are put into close places to sweat, which the Spaniards think makes them shear better) pass the summer in the elevated country, which supplies them with abundance of rich grass; and they do not leave the mountains till the frosts in September begin to damage the herbage.*

These travelling sheep are called by the Spaniards, *Trashumantes;* there is another breed of coarser woolled sheep, which are stationary, and called *Estantes*.

A flock in the aggregate, is called a Cavaña, and each distinct flock takes the name of its present or original possessor. The Cavaña of Paular, for instance, consisting of 36,000 sheep, originally belonged to the monks of Paular, from

---

* Sir Jos. Banks's Letter, p. 2.

from whom it was purchased by the Prince of Peace, soon after he rose into power, together with the land belonging to it in Estremadura and Leon, at a price per head equal to 16s. 8d. English.* The Cavaña of Negrete, takes its name from a title in the family of its present possessors; and that of the Escurial, from the monastery of that name, to whose monks it belongs.

By a code of laws, called the Mesta, every circumstance relating to the management of these sheep is conducted with the utmost precision.

Each cavaña is subdivided into tribes, of 1000 each; and the command of the cavaña, or whole flock, is confided to an officer called Mayoral, who generally possesses 4 or 500 sheep of his own, and is skilful in their management, and the cure of diseases. Each division of 1000 has five shepherds and four dogs to attend it: there are besides in each tribe about six Mansos, or

tame

---

* Sir Jos. Banks's Letter, p. 3.

tame wethers; these wear bells, and are very obedient to the voices of the shepherds, who generally feed them with bread; thus they follow, or are led by the shepherds, and the flock of course follows. The dogs are not used as in this country, for any purpose of driving or keeping the sheep together, but for their protection against wolves, which are very numerous, and while the sheep are travelling, very troublesome. At these times, and at the time of yeaning, each tribe is allowed one or two extra shepherds. So that according to the foregoing statements, these sheep in Spain are attended by thirty thousand shepherds and twenty-four thousand dogs; besides finding occasional employment for from five to ten thousand additional persons at the travelling and lambing seasons.*

Incredible as it may appear, it is nevertheless true, that by the laws of the Mesta, the flock is entirely confided to the care of the shepherds, without

---

* Sir Joseph Banks's Letter. Twiss's Travels in Spain.

without admitting any sort of interference by the proprietor, who derives no profit whatever except from wool, which after payment of duties, &c. is averaged at about one shilling per head.

A great deal has been said by different authors, and much importance has been annexed to the use of salt by the Merino sheep in Spain, but it was declared by the shepherds who lately came over with the Paular flock presented to his Majesty, that no salt is used, except in the very hottest season of the year, when the sheep are on the mountains.

These sheep, it is said, are always low kept;* indeed when the driving, sweating, and in short the whole of their treatment is considered, it does not appear to me possible how they should be otherwise than in low condition. As the number of the flock is prescribed, as well as the pastures which are to support them, all increase

* Sir Joseph Banks's Letter, p. 4.

crease more than is necessary to supply losses of their own, or a neighbouring flock, is useless; therefore most of the lambs are destroyed every year, as soon as dropped, and each of those preserved is made to suckle two ewes, by the practice (not at all unusual in this country) of putting the skin of the dead lamb upon the one which is to have the foster or additional mother, for a few days, by which means (the skin retaining the smell of her own lamb when alive) she very soon adopts it. The Spanish shepherds fancy, that the wool of an ewe which brings up her lamb without assistance, is injured.

At shearing time the shepherds, shearers, washers, and a multitude of unnecessary attendants are fed upon the flesh of the culled or drafted sheep; and it appears that the consumption at this season of feasting, is sufficient to devour the whole of the sheep that are drafted from the flock.* " Mutton in Spain," it is observed by Sir Joseph Banks, " is not a
favourite

---

* Sir Joseph Banks's Letter, p. 4.

favourite food; in truth, it is not in that country prepared for the palate as in this." It is however worthy of remark, that so partial to the Merino flavour were the shepherds who came over with the Paular flock, that on the road from Portsmouth, they cut up and eat every sheep which *died* upon the way, preferring it to the fine slaughtered meat of this country. The other breed of coarser woolled sheep, called *Estantes*, and the coast of Barbary, furnishes the supply that is wanted for the table.

When the Merino breed was first introduced into this country, an idea prevailed, that *not only* the fine air and herbage of the Spanish mountains, but that *travelling* was also necessary to promote the fineness of the wool. This idea has however been very successfully combated by Dr. Parry, by whom it is judiciously inferred, that the Estantes are not permitted to travel, *because they are coarse* and of less value; and *not* that they are coarse woolled because they do not travel.* We are informed indeed,

by

---

* Facts and Observations, p. 24.

by Bourgoing, that there are stationary Merino flocks, both in Leon and Estremadura, which produce wool *quite as fine as those which go to the mountains.* The sole purpose therefore of the journeys taken annually by these sheep, is to seek pasture where it can be found; hence it is clear that neither the journeys, nor mountain food, have any material effect upon their wool; which is also further proved by the experiments made in this and other northern countries, viz. Sweden, Denmark, &c. The adaption of the Merino breed of sheep to this country and climate, is therefore what we have next to consider.

In the year 1787, his Majesty, guided by those patriotic motives which are ever active in his mind, gave orders for the importation of Merino sheep for his own use, and for the improvement of British wool. As it was doubtful at that time whether the King of Spain's licence could be obtained, they were purchased on the confines of Portugal, and shipped at Lisbon. An application was afterwards made to the King of Spain, by Lord Auckland,

Auckland, the British minister at that court, and permission obtained to import some sheep drafted from one of the prima piles. His Majesty received, in consequence, in the year 1791, a small flock consisting of 36 ewes, four rams, and a manso. These were from the cavaña called Negrete, one of the three piles restricted from exportation, (as has been already observed) and which is likewise remarkable for producing the largest carcased sheep that are to be found among the Merino flocks. On their arrival, those that had been procured by way of Portugal, were all disposed of, in order that this breed might be kept *in its native purity,* which it still retains in every respect.*

The flock last imported, was a present to his Majesty from the government of Spain, and selected from the cavaña of Paular, of still greater estimation in Spain than the Negrete, and of which the exportation of its wool is also restricted. Two thousand were sent, but of

this

---

* Sir Joseph Banks's Letter, p. 8.

this number only about 1400 ewes and 100 rams were alive in February last. The ewes were heavy in lamb when they embarked, but owing to their very close confinement on board ship, and the want of proper food, having been only supplied with bad hay, several cast their lambs, and died at sea, and others immediately on their landing. The whole number of lambs in February did not exceed 350. From the state the animals were in on board the ship, by the account of the King's shepherd and others who saw them, and understand sheep, it is matter of surprize that so many have survived, under all the circumstances of their treatment. The shepherd declared that the heat and stench was so great that he was nearly suffocated in going into the ship's-hold. Sheep less hardy, and less used to heat and the confinement which these must have experienced in the *sudederos*, or sweating houses, would probably have died in much greater numbers. Of the survivors many were almost deprived of their wool, and were rendered very sickly by the severity of the season to which they were exposed immediately on

their

their landing, contrasted with the excessive heat they had experienced on ship-board. The scab also broke out very violently amongst them, probably from the same cause. Some have also died of the rot, which must have been contracted in Spain, as one died of that disor-der the day after they landed.

To return from this digression, the following statement of facts, relative to the Negrete flock of his Majesty, is so conclusive, that I am in-duced to quote the author's own words. This flock, it has been already stated, was imported in the year 1791.

" From that time to the present (says Sir Joseph Banks*) the opinion of the public, sometimes perhaps too unwary, and at others too cautious, in appreciating the value and adopting the use of novel kinds of sheep, has gradually inclined to give that preference to the Merinos which is so justly their due. At first it was impossible to find a purchaser willing to give
even

* Letter to Sir John Sinclair, p. 8.

even a moderate price either for the sheep, or for their wool; the shape of the sheep did not please the graziers, and the wool-staplers were utterly unable to judge of the merit of the wool, it being an article so many times finer and more valuable than any thing of the kind that had ever before passed through their hands. The butchers however were less timorous: they readily offered for the sheep, when fat, a fair mutton price; and there are two instances in which when the fat stock agreed for was exhausted, the butcher who had bought them anxiously enquired for more, because he said the mutton was so very much approved of by his best customers.

" It was not however till the year 1804, thirteen years after their first introduction, that it was deemed practicable to sell them by auction, the only certain means of placing animals in the hands of those persons who set the highest value upon them, and are consequently the most likely to take proper care of them. The attempt however succeeded; and the prices given demonstrated, that some at least of his

Majesty's

Majesty's subjects had at that time learned to put a due value on the benefit his royal patriotism offered to them. One of the rams sold at the first sale for 42 guineas, and two of the ewes for 11 guineas each; the average price at which the rams sold was 19*l*. 4*s*. and that of the ewes 8*l*. 15*s*. 6*d*. each.

" This most useful mode of distribution has since that time been annually continued, and the sales have taken place in the beginning of August. The last sale was held on the 17th of August, 1808, when the highest price given for a ram was 74*l*. 11*s*. 0*d*. for an ewe 38*l*. 17*s*. 0*d*. The average prices of rams was 33*l*. 10*s*. 1*d*. of ewes 23*l*. 12*s*. 5*d*.; a most decisive proof not only that the flock had risen materially in public estimation, but also that the sheep have not in any way degenerated from their original excellence.

" The wool was at first found to be quite as difficult of sale as the sheep themselves; manufacturers were therefore employed to make a considerable quantity of it into cloth, which,

when

when finished, was allowed by both woollen drapers and tailors to be quite as good as cloth made of wool imported from Spain. But even this proof would not satisfy the scruples of the wool buyers, or induce them to offer a price at all adequate to the real value of the article; it was found necessary, therefore, to have the wool scowered, and to sell it in that state as Spanish wool, which, though grown in England, it really was; thus managed, the sales were easily effected for some years, at a price equal to that demanded for the prima piles of imported Spanish wool at the times when the bargains were made.

" Time and patience have at last superseded all difficulties, and his Majesty's wool has now for some years been sold as clipped from the sheep's backs, the sheep having been washed, and the whole management of them carried on exactly in the English manner, at a price not lower than 4s. 6d. a pound, which allowing for the loss of weight in the scowering, costs the buyer at least 5s. 6d. a pound, a tolerable price for Spanish wool, when plenty of it could be

D produced,

produced, though not possibly so high a one as ought to have been given, or as will be obtained for the Anglo-Negrete pile, when the value of the article is fully understood."

This is the evidence of a gentleman of science and experience, than whom there is not a more honourable or disinterested character throughout the British dominions; and therefore, though I may again advert to the subject in a future chapter, I will not at present offend either the great authority I have quoted, or the common sense of my readers, by offering additional proofs on this head. That Merino wool does *not* degenerate in this country is sufficiently evident from the foregoing facts.

Now these sheep have taken no journey, eaten no salt, nor any of the fine herbage of the mountains of Leon and Estremadura, but have been *in every respect* treated in the English manner, and have had no more attention paid to them than to any other of his Majesty's sheep, of which he has considerable flocks. This I can state of my own knowledge, and Dr. Parry
tells

tells us, he has the authority of Sir Joseph Banks, who had then the direction of this flock, for saying, that "it has been fed on grass in the summer, and on hay in the winter; and that no particular management has been employed respecting the sheep, except that they have not been folded on fallow land, and that in winter and hot weather they have had access to a shed built on pillars, (without walls) under which they frequently lie down. And Sir Joseph adds, they have thriven full as well as other breeds of sheep kept on the same land, and under the care of the same shepherd."*

In respect to the production of fine wool in other climates than those of Spain, the following remarks of M. Lasteyrie, quoted by Lord Somerville, are so pertinent to the question, and so corroborative of the inferences adduced above, that I am induced to insert them here:

" The different Governments of Europe (he observes) had long acknowledged the advantages

D 2   that

---

* Dr. Parry's Facts and Observations, p. 28.

that would be derived to agriculture and com-. merce, from the introduction of fine wools into their respective states; but their views meeting opposition in the ignorance and prejudice of the times, a considerable number of years elapsed before they set about realizing an idea which at first seemed chimerical; at length there appeared men, equally commendable for their patriotism and their knowledge, who have laboured with zeal and perseverance, to enlighten their fellow citizens, by producing facts to prove that nature, far from opposing itself to the preservation of fine woolled sheep in certain climates, seemed on the contrary to lend itself complacently to the exertions of industry. I believe I have demonstrated in my treatise on sheep, that the fine wools of Spain *depend neither on the travelling, nor on the soil, nor the climate, nor the pasture*, but that they depend on other causes, and that it is possible to have in France, and elsewhere, wool of the same quality as that of Spain. My travels in the north of Europe have offered facts and observations, which have afresh demonstrated this truth. I have found in the far greater number of the flocks I have examined, wool, which, judging

from

from the eye or the touch, equals in beauty and fineness that of Segovia and Leon; so much so, that in my opinion no doubt can remain, that we can obtain superfine fleeces in every part of Europe, where pastures are to be found, and where we can depend on winter food, on which sheep can be supported. These wools make cloths as fine, as silky, and supple, as those manufactured of Spanish wool, as attempts made in France, and other countries, prove. But were it true, that the food, climate, and other local circumstances had a certain influence on the intrinsic qualities of wool, such as the elasticity, the strength, the softness, &c. it would not be the less proved, that at all events, cloths fine and beautiful enough to satisfy persons the most difficult on this point, can be obtained; and that a nation can easily do without the fine wools of Spain, and feed its finest manufactures with those drawn from its own proper soil."*

The Merino fleece is in colour unlike that of any English breed. There is a dark brown tinge

---

* Lord Somerville's Facts, p. 37, *et seq.*

tinge on the surface of the best fleeces, amounting almost to black, which is formed by dust adhering to the greasy yolky properties of its pile; the contrast between which, and the rich white colour within, creates great surprise on opening the wool. The harder the fleece is, and the more it resists any outward pressure of the hand, the more close and fine will be the wool.* This however, is not in *all* cases a positive rule, as Lord Somerville, at his last cattle show, produced several very fine rams, both for sale and hire, unexceptionable in point of wool, and of a very improved shape, most of which had *soft* and open fleeces. These I understand were descended from a favourite ram of his Lordship's, one of the most valuable ever produced in this kingdom, whose fleece was of the same nature.

There is occasionally a peculiarity in the wool of the lamb, when first dropped, differing from any breed in this country, (if indeed it may be termed *wool*) many of them appearing entirely covered

---

* Lord Somerville's Facts, p. 21.

covered with hair; which I do not find mentioned by any author I have met with except Dr. Parry, who observes that " the wool of the Merino lambs, in general, is evidently coarser and harder than that of the sheep. It seems, however, that different flocks vary in this respect. The lambs of the Infantado and Paular races are covered with a coarse sort of hair, which afterwards changes into very fine wool. The same appearance is sometimes to be found among the lambs of the Negrete breed in England."*

This sort of hairy covering, which appears also on many of the Anglo-Merino lambs, though certainly not ornamental, may be useful to the lamb, by keeping it warmer when first dropped than those which come more naked; and if so, it is perhaps a property rather to be wished than otherwise. Whether this circumstance is peculiar to any particular breeds, or from what natural cause it arises, I am entirely at a loss to determine. Only one out of six pure Merino ewes, of the Negrete breed,

---

* Communications to the Board of Ag. v. 5, p. 346.

breed, which I purchased at the King's sale in the year 1807, produced a lamb so covered, this year; but I had a great many yeaned so from my Anglo-Merinos, and as several of these are descended from Lord Somerville's rams which he procured in Spain, I am not sufficiently acquainted with their pedigrees to know from what flock they are derived. It is however, not a matter to be regretted, for I have in general found that all the hair falls off in two or three months, and is succeeded by wool of the finest quality.

I was fortunate enough to procure last year, from Kew, some three-quarter-bred Merino-Ryeland ewes, which were put to one of his Majesty's rams. Three of them produced red lambs, which at first had much more the appearance of foxes than lambs, being precisely of the colour of the former: they were all ewes, and exactly alike, with a white tag at the end of their tails. I was extremely uneasy at this circumstance, as their mothers were some of the very best woolled and best shaped sheep I have, least some West India or other foreign ram had

by

by any accident been with them, for the native West India sheep are of that colour; but when they were a month old, I was quite satisfied the hair would come off; and though not yet three months old, they are now nearly white, and as far as the hair has fallen off, it is succeeded by very beautiful white wool, and of a quality equal, or superior to that of their dams.* This has already taken place on more than half the carcase, and it is nearly the same in all the three. I must confess that this hair very much disfigures the lamb while it remains in the state of half fine wool, and half long hair, even when white: and some of these, I fancy, are to be found in every Merino and Anglo-Merino flock. I have stated this circumstance, as I think it my duty to mention every thing I know about these sheep, whether to their advantage or not. My object is to inform, and not to mislead, and to prevent any beginners from being discouraged or taking objections to these breeds, on finding any lambs

in

---

* Samples of this hair and wool may be seen at the publisher's.

in the early part of the season, in this state;
for if they will examine the same flocks a few
months afterwards, they will perceive this objectionable appearance entirely removed. And
as it is difficult, when a gentleman or farmer is
himself satisfied with, and determined to try any
new breed of sheep, or system of husbandry,
to prevail on his bailiff and servants, or shepherd, to adopt his ideas, and abandon their old
prejudices, it is better to apprize them of such
circumstances as these, otherwise their prejudices may be confirmed and increased, and
they may be as much surprised at a novel
appearance as my shepherd's boy was in the
winter, who declared when he found the red
lambs in the morning, that two of the ewes
had yeaned young foxes. Such stories as
these being disseminated are well calculated
to prejudice others, as has been the case
with a Mr. John Hunt, of Loughborough,
(a gentleman well known to the readers of the
Agricultural Magazine) who having heard of
one of these hairy lambs, mentions it with great
exultation as a proof of the degeneracy of Merino sheep in this climate. This gentleman,

who

who has occupied many pages in several numbers of the above-mentioned publication, with the most violent declamations in praise of the Dishley, or new Leicester sheep, and in abuse of the Merinos, has certainly done the Merino cause more service than I dare venture to hope my poor abilities will ever be able to do; for he has brought them forward to the public, loaded with every objection, imputation, and fault, that either ingenuity can invent, or ignorance assert; and by these means has given persons better acquainted with them an opportunity of answering him most completely, and of shewing the inaccuracy, folly, and untruth of all his statements and suppositions, controverting them by facts, supported by names and dates. He has also afforded the readers of the Agricultural Magazine the means of comparing the Merino with the Dishley sheep, and learning the advantages of crosses with them, which might not otherwise have occurred. Amongst the gentlemen who have taken the trouble of refuting his statements, Benjamin Thompson, Esq. of Red Hill Lodge, near Nottingham, stands the most conspicuous, and these answers are not only replete

plete with able and convincing argument in reply to all Mr. J. Hunt's rhapsodical and ridiculous ssaertions, but contain much other amusing and instructing matter. Another gentleman, who signs himself Cultivator, has also taken the trouble of replying to Mr. J. Hunt in the same work, to which I beg leave to refer my readers, as further mention of him in these pages will far exceed my limits. And but for his absurd remark on the hairy lamb, and my fear that he may suppose I have overlooked his labours, I should not have noticed him at all, for when any person asserts, as Mr. John Hunt has done, disdaining all proofs, or the offer of any, and in the teeth of the most unquestionable authority, and of several years experience in this country and in France, and the more northern countries of Holland, Sweden, and Denmark, communicated by men of the utmost integrity and eminence,* that " *if a male and female Merino sheep were brought into Leicestershire, in a few years the nature of their offspring would become subservient*

---

* Namely, Lord Somerville, Sir Joseph Banks, Dr. Parry, M. Lasteyrie, M. Pictet, &c.

subservient to local circumstances, were even no crossing to take place; the carcase would improve, and the wool become coarse," and that " *if a male and female Leicestershire sheep were taken into Spain, the wool would in time become similar to the natural production of the country,"** I do not think he is entitled to any answer, for if he had said that such exchange of situation would make them goats, the assertion would be equally probable. I do not know what obligations the Leicestershire breeders may feel themselves under to their Loughborough champion, but I am sure the Merino breeders are very much indebted to him for the additional publicity he has caused of the merits of the latter.

I shall now proceed to state such particulars as are known of the health and longevity of this race, properties which encourage us to possess *the pure blood*, when it can be obtained, even at very high prices; or to approach as near to it as possible, for it appears as far as I can learn from others, and from near ten years experience of my own, that the descendants inherit these

---

* Agricultural Mag. Aug. 1808,

these invaluable qualities in proportion to their consanguinity to their pure Merino ancestors.

From the few years the Merinos have been in this country, it is not likely a very great deal can be known from experience of those that have been bred here; and from their being so abundant in Spain as to make it necessary to destroy the lambs, it is not more likely any are allowed to live after they become broken mouthed, or longer, probably, than to 7 or 8 years of age, therefore no information will probably arise from thence. In countries where they are so difficult to procure as in France, and in this and other northern countries into which they have been introduced, it has been and is still a great object to keep them alive as long as they are capable of breeding. Monsieur C. Pictet, of Geneva, says, that the Merino is longer coming to maturity than most other breeds, and that they do not acquire their full growth till three years old; and that they shed and renew their teeth some months later than the native breeds of France.*

If

---

* Pictet, Faits et observations sur les Merinos d'Espagne, quoted by Dr. Parry in his Essay, p. 347.

If however, as is stated above, these sheep
are slower in becoming adult, it is generally
agreed that they are much longer lived than
any other known races. They sometimes keep
their teeth to 14 or 15 years of age, and
according to M. Pictet, there was in the
possession of Citizen Marais, at Nogent, in
the year 1802, a Merino ewe, which, having
come from Spain in the year 1786, could not be
less than 16 years old: she had then all her
teeth, and had brought a lamb the preceding
winter. I have myself three Merino ewes certainly very old, but as two of them had no
teeth when I bought them, and those who sold
them to me did not breed them or know their
age, I can only give the following calculation,
which I think satisfactorily proves one of them
to be now at least 12 or 13 years old; and
they are now much better, and shew less
appearance of decay, than when I purchased
them, probably from having been better attended to.

A sheep does not become full mouthed till
five years old, and unless fed with Swedish turnips

nips, or very hard food, will retain her teeth whole, and not become what is called broken mouthed till two or three years after, which would be eight years. They will then probably retain several of their teeth, unless they are pulled out, at least two years more, which amounts to ten years. I purchased one of these ewes in lamb at Christmas 1805, when having no teeth she was probably ten years old. I think, therefore, the fair probability is, that she is now twelve or thirteen years old at least. She brought up her lamb in 1806, and has done so ever since till this year, when I put her lamb to another ewe. She is in full health, as are also the other two, which last I purchased at the King's sale in 1807: one of these had then no teeth, the other had a few ragged and loose ones, which I ordered to be immediately pulled out. They both brought very fine lambs in 1808, and this year produced twins, three fine ram lambs, and an ewe, which are all alive.

These ewes, if attended to by being kept in rouen and long grass, or having their food cut, keep themselves in full as good condition as any

of

of my flock. I do not intend to let them breed up any more of their own lambs, but put them to a wet nurse or foster mother as soon after they are dropped as I can procure one, which prevents the ewe from getting poor, as suckling would of course make her. Besides, the lambs being put to younger ewes full of juices and good milk, have a much better chance of being well brought up; and with this sort of attention I have no doubt the ewes will yet live some time. Of seven pure Merino ewes which I have purchased of his Majesty, I have lost but one, though they were all old, and some of them the worst of the sale in 1807. That which died, never recovered her journey to my house, being in a very weak state when bought, though she had all her teeth and was sound.

M. Pictet mentions some curious particulars with regard to the Merino race, as for example, " that they eat more indifferently of all sorts of food than other sheep; and that the adult sheep have an erect mien and measured step, in which respect they seem to partake of the state-liness and gravity which characterize the human

E   inhabitants

inhabitants of their native country; that they are also remarkably timid; and that nothing seems to give them activity but fear, hunger, sensual desire, or jealousy." This I have always observed of them and their descendants, that they are remarkably *quiet* and satisfied in their pastures, and not disposed to break or creep out, or to jump over hurdles. It seems to be universally agreed, that the rams of this breed are extremely salacious.

In concluding this chapter, I beg to call the reader's attention to the improvement in shape of which the Merino sheep is susceptible, and in so doing, I cannot urge any thing more forcible than the following language of my Lord Somerville:

" No attention, (says his Lordship) has been paid in Spain to the form of their sheep; and it must be evident to every judge of stock, that a journey from the mountains of the north, to the plains of the south of Spain, cannot be otherwise than productive of more injury to the frame and constitution of the animal than of benefit

benefit to the fleece, which; like the frame, is dependant on, and nourished by the blood. Does it stand to reason, that a long drift of four or five hundred miles every spring and autumn, and that at the rate of 80 or 100 miles per week, can be beneficial to sheep? Have we a single breed which could support it uninjured? None which would not be hunted into deformity."

There is in the pure Merino frequently a pendulous skin under the throat, what we term throatiness, much esteemed in Spain because it is supposed to denote a tendency both to wool and to a heavy fleece. This property is usually accompanied with a sinking or hollow in the neck, both which are extremely offensive to the eye of an English breeder; but a judicious drafting of the faulty ewes, and a due attention to the form of the rams, will in a few years remove all traces of these deformities, for there is no reasonable foundation to conclude that a bad shape is essential to the production of good wool. The fact is not so. To make up in some degree for the faulty points just mentioned, these sheep

E 2 are

are in general very straight in the back and
behind the shoulders.*

The attainment of a fine carcase in the New
Leicester and South Down sheep, it is justly
remarked by Dr. Parry, has required many
years of attentive study. " It has been effected, (he continues) among other means, by a
selection of the best made animals for the purpose of breeding; and we cannot doubt that the
same care applied to the Spanish race will produce the same effect."†

CHAP

---

* Lord Somerville's Facts and Observations, pp. 22, 24.
† Dr. Parry's Facts and Observations, p. 42.

CHAP. III.

On the Anglo-Merinos, containing observations on the different crosses, and an account of the success which has hitherto attended the exertions of the English breeder in improving the fleeces of our native sheep, by admixture with the Spanish; in which are demonstrated the advantages resulting to the practical farmer from the use of the Merino and Anglo-Merino sheep, compared with other breeds. And also on the improvement of the Merino carcase by such crosses; with remarks and facts, tending to prove the natural disposition to fatten in the Anglo-Merinos, and that they are a very healthy and hardy race; with unquestionable evidence as to the excellence of their mutton.

OF our native English breeds of sheep, the Ryeland and South-Down produce the finest clothing wool, and are on that account the best

adapted

adapted for crossing with the Spanish. The Ryeland is a Herefordshire sheep, and has long been generally acknowledged to produce the finest wool of any of our native breeds; but having been of late years crossed with rams of the new Leicester or Dishley breed, from the prevailing rage for that race, the wool has been much deteriorated, and it is consequently become very difficult to procure the pure original stock. They are a polled, or hornless sheep, originally inhabiting, not as has been supposed, a mountainous country, but the vales and lands about Hereford, Ross, and Ledbury, the same lands which produce the fine Herefordshire oxen.* They produce from 1¼ lb. to 2¼ lbs. of clothing wool, of which a small part about the neck and shoulders is often very fine. There is, however, a horned black-faced breed of sheep, called Morfe, bred on the commons about Bridgnorth, in Shropshire, which is said to produce still finer wool than the Ryeland, though in smaller quantity. The filaments of these two

sorts

---

* Lord Somerville's Facts, *p.* 14.

sorts of wool are nearly as small as those of the finest Merino; but they are more irregular in size and surface, and consequently rougher. Neither does this wool felt well; that is, it is intractable, and does not yield to the proper pressure, and retain its form so as to thicken as it ought to do on its manufacture into cloth.* The Morfe, too, being a heath sheep, of a wilder nature, and not so docile as the Ryeland, is found by experiment not to be so desirable for crossing with as the latter.

The South-Down, as a cross with the Merino, has its advocates, and there are many very valuable flocks of this admixture in England, which I think in some points exceeds the Ryeland; but I agree entirely with Lord Somerville, Dr. Parry, and others, in opinion that the Ryeland cross has the superiority in fineness of wool, and certainly in that fine, soft, and silky quality so peculiar to itself.

On crossing our native breeds with Spanish rams,

---

* Dr. Parry's Facts, *p.* 3.

rams, the progeny partake of the blood of the
sire in the following proportions:

The produce of the first cross will be half blood;
   or 50 parts Merino out of 100.
The second cross gives ¾ths Merino, or 75 } parts
The third . . . ⅞ths . . . or 87½ } out of
The fourth . . 15-16ths . . . or 93¾ } 100.

And after the fourth cross, the wool of the Ryeland or South-Down admixtures is so much
ameliorated, as not to be distinguishable from
that of the pure blood. This close approxima-
tion to the Spanish after the fourth cross, it may
be remarked, is peculiar to the Ryeland and
South-Down admixtures; in most other breeds
it requires more crossing to produce wool of the
same quality. Dr. Parry says, " Experience
shews that a fourth cross with a Ryeland ewe
gives the specific predisposition, which is capable of being communicated to posterity, as well
as from the pure Merino breed. At the same
time, as the individual constitution is also not
without some capacity of being transmitted, a
Merino-Ryeland ram of the fourth cross, with
                                        finer

finer wool, (under similar states of health and flesh,) than the original breed, will, by the union of these qualities, give his posterity finer fleeces than that breed itself: and this effect may with certainty be relied on, if it shall have been found that three or four generations from such a ram shall have maintained this superiority of fleece. If we stop at two or three crosses, we may have rams, of which some individuals may have wool as fine as the Merino; but as this may arise from accidental or acquired constitution; that it will perpetuate itself with ewes of precisely the same mixture is not to be relied upon."* After the fourth cross, fine wool being obtained, the Doctor recommends breeding in and in.

The quantity of wool, too, is as proportionably increased by the Merino cross, as the quality is improved thereby. The Ryeland sheep is stated to carry a fleece of about 2½lbs. The Merino-Ryeland ewes of the second and third crosses

---

* Bath Papers, *vol.* 11, *p.* 318.

crosses will often produce 4½ and 5 lbs. and on the average of a flock at least 4 lbs. The fourth cross will average at least 5 lbs. As an instance of what may be obtained by care and attention, it may here be stated, that ten Merino-Ryeland rams, of different ages, from four-teeth upwards, purchased of Dr. Parry, by Mr. Birkbeck, yielded in 1806, 97 lbs. of wool, or nearly 9¾ lbs. each, in the yolk; and of these, two gave 11 lbs. 15 oz. each, and one, which had the finest fleece, 12 lbs.*

In the fleeces of the Spanish race, and also in those of the mixed blood, the proportion of fine wool is much greater than in those of any native English breed. The finest wool of the English breeds, on a proper division of the fleece by the woolstapler, does not produce, *of the finest parts*, above *one-eighth* of the whole fleece; while nearly *four parts out of five* of the Merino fleece, are equally fine. The proportions of the different sorts of wool, it may be observed, vary a little in different

---

* Communications to the Board of Agriculture, *vol.* v. *p.* 537.

different fleeces. The finer the whole fleece, the greater the relative quantity of the finer parts.*

Another valuable peculiarity of the Merinos, is this:—The wool of all the English breeds is finest on the shoulder, and in general coarse, at least always *coarsest* on the breech or hams; while in the Merinos and the mixed blood, the wool on both those parts is equally fine, or very nearly so. Dr. Parry, on this subject, speaking of his Merino-Ryeland breed, makes the following observation:—" The uniformity of fineness in different parts of the fleece of my breed of sheep, is what I fear I shall not readily make some persons believe, on my bare assertion. The fact, however, is, that in shewing the specimens cut by myself from those different parts of the same animal, which are generally considered as producing the best and worst wool, I mean the shoulder and the breech, I have never met with three persons who could agree

---

* Dr. Parry. Communications to the Board of Agriculture, *Vol.* v, *p.* 444.

agree which was the finest. Many good judges have actually decided in favour of the latter. This is a curious and important distinction between the Spanish and English breeds."*

To this observation made by the Doctor, when the breed was here in its infancy, I have now to add my testimony, as will, I have no doubt, every gentleman who has at all attended to the subject. So remarkable is this circumstance, that was an Anglo-Merino to be put into a flock of 100 sheep, similar in general appearance, any person at all skilled in wool, or in the breed, would immediately point it out by the closeness of its wool on those parts, which in other sheep are usually loose and hairy.

In order to the subject's being better understood by those at present unacquainted with it, it will be necessary before I proceed farther, to shew the manner in which the Spaniards divide their wool, and the terms by which the separate parts

* Dr. Parry's Facts and Observations, p. 8.

parts are distinguished. The fleece is divided
into four parts, the first of which is called
Refina; the second, *Fina;* the third, *Tercera;*
the fourth, *Cahidas.* The parts of the sheep
on which these respective sorts grow, will be
more fully illustrated by the figures on the following drawing.

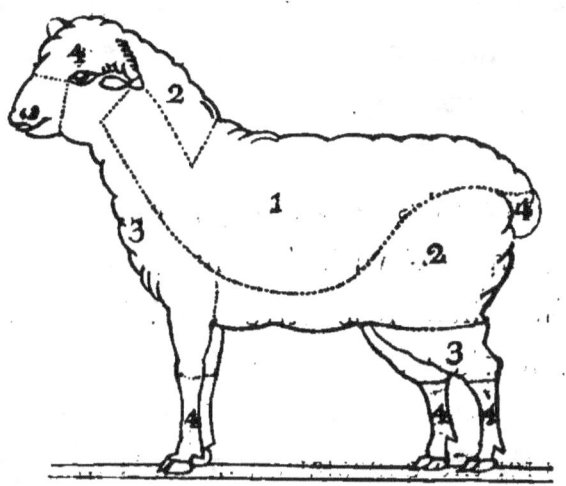

No. 1. The *Refina*—is found on the flanks,
the back as far as the tail, the shoulders, and
the sides of the neck.

No. 2. The *Fina*—on the top of the neck,
the haunches, as far as the line of the belly, and
the belly itself.

No.

No. 3. The *Tercera*—on the jaws, the throat, the breast, the fore thighs to the knees, and the hinder thighs from the line of the belly down to the hocks.

No. 4. The *Cahidas*—is that which grows below the hocks, between the thighs, on the tail, the buttocks, the pole, and behind the ears; and all that which shakes out of the fleece in shearing, or in washing.

The three first sorts, only, come to England, in bags marked with the initials R. F. and T. denoting their respective qualities. The Cahidas, marked with C or K are sold in Spain, the small amount of which is allotted to pay for masses for the consolation of souls in purgatory. In this country, what is produced, will no doubt be applied to at least as beneficial a purpose.

The amelioration which has been effected in the fleeces of such of our native breeds as have been crossed with the Spanish, is next to be considered.

In

In the year 1799, the Bath and West of England Agricultural Society offered the two following premiums:

" To the manufacturer of the finest piece of cloth, made from English wool, a navy blue, dyed in the wool, not less than 25 yards, to be produced at or before the September meeting 1800; Plate, value ten guineas.

" To the manufacturer of the finest piece of white kerseymere, made from English wool, not less than 25 yards, to be produced at or before the September meeting 1800; Plate, value five guineas."

Three pieces of cloth, and two of cassimere were produced for these premiums. Two of the pieces of cloth were made from the best Ryeland wool; but the premium was adjudged without hesitation, by the committee of manufacturers, to Mr. Joyce, of Freshford, for the third piece, which was made from the wool of Dr. Parry's Merino-Ryeland breed. The Doctor

gives

gives the following account of it. It was di-vided by a woolstapler at his house, into three sorts, in the Spanish method. The fleeces weighed in a raw state 155½ lbs. of which the Refina made 104 lbs. the Fina 38½ lbs. and the Tercera 13 lbs. Out of the 104 lbs. 92½ lbs. were selected and sent to Mr. Joyce. This wool, when scowered, weighed 50 lbs.; when woaded blue, and picked, 48 lbs. The warp ran 22 skains to the pound; and the abb, or woof, 18 skains. When set on the loom it was in length 31½ yards. When finished, it made 29½ yards of cloth, worth 19 or 20 shillings per yard, the price at that time of the superfine broad cloth.

It further appears that this wool was, from the ignorance of the sorter, very ill-sorted; that Mr. Joyce used every lock of it; and that, compared with the finer wool belonging to Dr. Parry, and taken from his flock only, it was fit only to be estimated in the second degree, or Fina. Notwithstanding all this, the cloth was computed to be worth from three to four shil-lings

lings per yard more than the other two pieces, made of Ryeland wool, its competitors.*

The prize piece of cassimere was made by Mr. Waldron, of Trowbridge, also from Dr. Parry's wool, selected by himself and one of his servants, from the fleeces of shearling rams and ewes. The whole was worked into yarn. It weighed raw 36¼lbs; scowered and dry 19¼lbs; and made 19 lbs. 2 oz. of yarn. Of this, 17 lbs. 14 oz. were woven, and produced, of uniform cassimere, 32¼ yards, which weighed 15¼lbs. This cassimere was equal to any made from the best Spanish wool indiscriminately used.†

On an examination of these cloths, the committee were unanimously of opinion that no chance of success could hereafter attend any cloths manufactured with the finest Ryeland wool, in competition with those made from this mixed breed. The manufacturers allowed that the

F   yarn

* Dr. Parry's Facts. p. 33.   † Ibid. p. 34.

yarn was free from what in their technical phrase is called nibs.

And this wool, Dr. Parry adds, was grown under the indiscriminate use of the coarsest food, in small inclosures, without housing, or any other management than what is common to the hardiest and most ordinary of our sheep.*

Now these facts are decisive of the amelioration that has taken place within a few years, in the fleece of the Ryeland breed, on being crossed with the Merino. Let us next proceed to consider the result of experiments that have been made, in comparison with the best imported Spanish wool.

In January, 1804, Lord Somerville sent to the Bath Society a pattern of some broad cloth, the produce of 10 fleeces of wool, quantity 14¼ yards, that the society might form some judgment

---

* Dr. Parry's Facts, p. 35, 37.

ment of the average value of his Merino wool, as applied to the fabric of broad cloths. This cloth, on a minute inspection of the committee and eminent manufacturers, particularly Mr. Naish, of Twiverton, was allowed to be worth 20s. per yard, and only to be inferior to the *very best* Spanish cloth 2s. per yard. But it was admitted that some allowance should be made for the injury the pattern had sustained by carriage, and in passing through so many hands in its examination, which had injured its original appearance. And it was also admitted, that the wool was *susceptible of a better manufacture than it had received,* to the amount of probably 1s. or 1s. 6d. per yard.* The following remark by the editor of the Bath papers on this report, is particularly worthy of attention. " The foregoing account of Lord Somerville's success in manufacture is not inserted as one of the *most* flattering experiments, but rather one of the *least.* The pattern of cloth alluded to, though substantial and good, is certainly inferior to

F 2   cloth

---

* Bath Agr. Society's Papers, vol. 10, p. 67—71

cloth since manufactured from his Lordship's wool, from other wool produced from Dr. Parry's flock, from the Secretary's, and from sundry other breeds of the Anglo-Spanish sheep. Indeed so great is the improvement of such wool now become, that cloth made from it readily obtains amongst good judges the character of being equal to almost any made from imported Spanish wool, and superior to the general fabrics."

In November, 1806, Dr. Parry obtained two other premiums of the same Society, at the annual meeting of which, the Committee appointed to examine cloth and wool, reported as follows:

" That they had, in conjunction with several woollen-drapers of the city of Bath, minutely inspected the comparative quality of cloths, Nos. 1 and 2, and had adjudged that No. 1 *was entitled to a preference in respect to fineness of wool.* It appeared that No. 1 was manufactured from wool of Dr. Parry's Anglo-Merino breed; and that No. 2 was made from one of the best piles of wool imported from Spain, and known by the

the name of the Coronet pile. They had also examined a piece of navy blue British cloth, manufactured by Mr. Joyce from wool of Lord Somerville; also a piece of white uniform kerseymere from the wool of Dr. Parry; both which they declared to be of excellent quality, both as to fineness of wool and firmness of fabric, and *equal in every respect* to the generality of cloths made with the best wool imported from Spain. On the whole, they were decidedly of opinion, that Dr. Parry had, by his zeal, diligence, perseverance, and activity, accomplished the grand object of producing *in the climate and soil of Britain*, wool equal to that usually imported from Spain."*

Signed,

J. BILLINGSLEY.

The cloths, No. 1 and 2, were both manufactured by Messrs. Yeats and Son, of Monk's Mill, near Wotton under Edge. Their foreman,

to

---

* Bath Papers, *Vol.* xi. *p.* 163, 164.

to whose superintendance the making of the two cloths was entrusted, remarked, that Dr. Parry's wool and cloth in every state worked much more kindly than the Coronet.

It is stated by Dr. Parry, in a letter to the President and Members of the Society, that the wool of the cloth made by Messrs. Yeats and Son, in comparison with that made from the Refina wool of the celebrated Coronet pile from Spain, consisted of ewes' fleeces from his flock *descended from Ryeland ewes*, crossed with rams from the King's and Lord Somerville's Merino flocks, to the fourth generation. That the British cloth would have been much finer but for the unskilfulness of the wool-sorter, who, notwithstanding he had his choice of a great many better fleeces, admitted several of a coarser kind. That the same observation applies to the wool of which the cassimere was manufactured by Mr. Joyce, comparatively with a piece made from the native Spanish pile marked R.X.S.*

The

---

* Bath Papers, *p.* 165.

The Doctor very handsomely waved his claim to the premium for both these articles, being perfectly satisfied with the concurrence of the Committee, and the manufacturers of both cloths, in the opinion which he had long entertained, that the wool of the Merino-Ryeland breed, properly cultivated in Great Britain, will make cloth and cassimere at least equal in quality to the best wool of the pure Merino race.

To the nation at large, as well as to the practical farmer, these facts are of the utmost importance. They sufficiently demonstrate *not only* that fine wool can be produced in this country, adapted to the manufacture of the finest broad cloth, but also that *a finer staple* has been actually obtained, from an improved breed of sheep, and that it has made *better* cloth than the best samples imported from Spain.

Now when the management of the Merino sheep in Spain, as has been already detailed in the preceding chapter, is considered, and the very confined knowledge of the persons to whom

that

that management is entrusted, and the limited means possessed by the Spaniards of attempting improvements, even if they had the disposition, bound down as they are by their ancient customs and the laws of the Mesta, improvement in Spain seems out of the question, and we must therefore look to the enterprising spirit of this country for such amelioration, either of wool or carcase, as the Merino sheep are susceptible of. The great and long prevailing objection of the unfavourableness of our climate being now removed, the knowledge and attention of English breeders cannot fail to effect great improvements in both these points. In this opinion I am warranted by the successful result of the experiments that have been already made. It has been observed by Dr. Parry, that but for the silly prepossession which formerly prevailed not only in England but all over Europe, that fine Merino wool could only be produced in Spain, the Spaniards would never have suffered a single Merino to have been exported.

In addition to the evidence which has already been adduced, the superior fineness of the Merino-Ryeland

rino-Ryeland wool has been ascertained by a *still more incontrovertible test*, that of actual microscopic admeasurement; the process of which, and the means adopted by Dr. Parry for ascertaining the experiment, are as curious as they are interesting. The reader will be amply gratified by referring to the Doctor's Essay (already so repeatedly quoted in these pages) for farther illustration.

This wool, too, suffers less waste in the manufacture than the native Spanish, which is accounted for by Dr. Parry on the following just principles:—" The Merino sheep in Spain remain at rest during a certain time in the rich vallies; and, from the luxuriance of food, become tolerably fat. They are then reduced in flesh by a long journey to the mountains, where they rest, again increase in obesity, and are once more reduced by travelling to be again fattened by repose among the rich herbage of the lowlands. These opposite states of rest and motion, accompanied as they are with a corresponding degree of nutrition, must be very unfavourable to the equal growth of the wool,
and,

and, therefore, to the tenacity and uniform strength of the filament; while, on the other hand, a well-managed flock of sheep in this country, kept in nearly an equal state of exercise and flesh during different seasons of the year, must furnish wool of more uniform strength, and therefore less subject to waste in the different processes of the manufacture."*

It farther appears from Dr. Parry's Letter to the Bath Society, that in the last instance the sheep producing the prize wool were kept in excellent order for a full twelvemonth before shearing, having been fed in the respective seasons not only with grass and hay, but with vetches, clover, cabbages, potatoes, linseed, and oil cake; and some judgment (he adds) may be formed as to their healthiness, when it is known, that from the time of ramming in September, to that of shearing in June following, out of 102 ewes only three died, two of which were killed by scouring, after having gorged themselves with boiled potatoes mixed with salt, which being

new

---

* Communications to the Board of Agriculture, *Vol.* v. *p.* 453.

new to the flock, much affected those even which recovered.*

But to return to the preceding facts. They are of considerable importance to the practical farmer, inasmuch as they evince the state of perfection to which the Anglo-Merinos have been brought in respect to wool, the comparative value of which with that of other breeds, is next to be considered.

From an account procured by Lord Somerville, at the request of Dr. Parry, from Arthur Young, Esq. the following appears to be the average weight of the fleece relatively to the living weight of an ewe in tolerable flesh, of the respective breeds :†

	lbs.		lbs.
Lincoln fleece	8¼	living ewe	140
New Leicester	7	. . .	130
South-Down	3	. . .	125

To

---

* Bath Papers, *Vol.* xi. *p.* 165.

† Dr. Parry's Facts, *p.* 44.

To which Dr. Parry adds,

	lbs.		lbs.
Ryeland	$1\frac{3}{4}$	Living ewe	60
Merino-Ryeland (washed)	4	. . . . .	60

Whence it appears, that the wool is to the living weight,

		lb.		lbs.
In the Lincoln	as	1	to	$16\frac{1}{2}$
New Leicester	-	1	to	$18\frac{1}{2}$
South-Down	-	1	to	$41\frac{2}{3}$
Ryeland	-	1	to	$34\frac{2}{7}$
Merino-Ryeland	-	1	to	15

And the value of the wool (in 1800),

	lb.		s.	d.		s.	d.
Lincolnshire . .	$8\frac{1}{2}$	at	0	$8\frac{1}{2}$	per lb. is	6	$0\frac{1}{4}$
New Leicester .	7	-	0	$8\frac{3}{4}$	. . . .	5	$1\frac{1}{4}$
South-Down . .	3	-	1	10	. . . .	5	6
Ryeland	$1\frac{3}{4}$	-	2	4	. . . .	4	1
Merino-Ryeland (unwashed)	$4\frac{1}{2}$	-	2	10*	. . . .	12	9

In

* Compared with the then price of Spanish Refina, this Merino-Ryeland wool might fairly have been reckoned at 3s. $1\frac{1}{2}$d. per lb.—*Dr. Parry's Facts, p. 46.*

In order to shew the difference of value still more clearly, the Doctor then states the value of the wool of each different breed proportionably to one given living weight of carcase. If we bring them all to 140 lbs. the stated weight of the Lincoln, then the account will stand as follows:

Annual Produce of wool on 140 lbs. of living weight.

	£.	s.	d.
In the Lincoln breed	0	6	0¼
New Leicester, about	0	5	6
South-Down	0	6	1¾
Ryeland	0	9	6¼
Merino-Ryeland . . .	1	9	9

Hence it appears, that the Merino-Ryeland breed carries more than three times the value of wool on the same living weight of carcase that its Ryeland ancestor does; almost five times as much as the South-Down and Lincoln; and nearly five and a half times as much as the New Leicester.*

The value of the Anglo-Merino breed is farther shewn by the following account, extracted from a communication of Edward Sheppard, Esq.

of

---

* Dr. Parry's Facts, p. 46.

of Uley, in Gloucestershire, a very considerable manufacturer, and great Merino proprietor, to Sir John Sinclair. Mr. Sheppard remarks, that the weight of the Ryeland fleece was increased one half by the first cross with the Spaniard. In 1806, his mixed flock, amounting to 986, averaged upwards of 3 lbs. each fleece (washed on the sheep's back) producing, at the then low price of wool, the sum of £676. 1. 0. Thus, 986 fleeces, weighing 3,033 lbs. averaged nearly 4s. 6d. per lb.*

It will be perceived by the reader, that the foregoing calculations, as well as the preceding facts, apply exclusively to the Anglo-Merinos of the Ryeland cross. In the following statement of the return of wool per acre, published by Lord Somerville, in 1803, is shewn the comparative value of the South-Down, as well as the Ryeland admixtures, of the first cross, with the fleece of the parent stock: and from this statement may also be collected some idea of the value of the South-Down, and Ryeland crosses, in relation to each other.

Wool

---

* Communications to the Board of Agriculture, *Vol.* vi. *p.* 66.

Wool Produce per Acre.

South-Down store ewes, at 3lbs. a fleece, and
 at 1s. 10d. per lb. will pay 5s. 6d. per
 fleece; which, at 6½ sheep per acre, in good
 upland pasture for seven months, and five
 months in turnips at the rate of 14 or 15
 per acre, will pay per acre, 1l. 18s. or   2 0 0

Ryeland ditto, at 2½lbs. each fleece, and at
 2s. 2d. per lb. untrinded, will pay 4s. 10½d.
 per fleece: and 9 sheep per acre, turnips
 in the same proportion as above, will pay
 per acre 2 3 10½

South-Down and Merino ewes, of the half-
 blood, at 4lbs. a fleece, clean washed, and
 at 3s. per lb. will amount to 12s. each
 fleece; and 7½ sheep per acre, with turnips
 for winter keep, will pay per acre . . . 4 10 0

Merino-Ryeland ditto, half-blood, at 3₂lbs.
 per fleece, and 3s. 2d. per lb. will pay
 10s. 3½d. each fleece, and 10 sheep per
 acre, with turnips in proportion as above,
 will amount to 5 2 11*

            " Doubts

---

* In the original work, from which this calculation is
extracted, this amount is given at £6: 10s. 5d. which
is evidently an error of the press, as will be perceived
by the reader on calculating the produce of 10 sheep, as
stated per acre, at 10s. 3½d. each fleece.

" Doubts (says his Lordship) have arisen as to the possibility of any return, such as this, to be had from wool. What then will be said of the pure Merinos? Why, that their fleeces have never been sold at less than one guinea each; that the weight averages more than 6 lbs. each, in the yolk; that on the above allowance of pasture for seven months, and turnips in aid of that pasture, the return shall be 10 guineas per acre.*"

In the account quoted of Mr. Tollett's pure Merinos, in a preceding page,† it is shewn that his flock averaged in 1804, 6 lbs. 6 oz. each, of wool in the yolk, which sold for 3s. 1½d. per lb. through all the sorts. Dr. Parry informs us that his Merino-Ryelands averaged in 1806, upwards of 5 lbs. each in the yolk, worth 3s. 9d. a lb.‡ An instance has already been adduced in page 58 of this Treatise, of ten of the Doctor's

---

* Lord Somerville's Facts, p. 59.

† Vide page 17.

‡ Communications to Board of Agriculture, vol. 5, pp. 446, 451, 537.

tor's rams, sold to Mr. Birkbeck, of Wanborough, yielding nearly 9¾lbs. each. The *washed* wool of the King's flock has been stated, on the authority of Sir Joseph Banks, to have sold at not less than 4*s.* 6*d.* per lb.*

Now these facts, in addition to what I have before quoted, are too obvious to need any comment. The increased profit to be derived from the wool produce of a Merino, or Anglo-Merino flock, is indisputably demonstrated.

It is to be observed that the preceding calculations were all made when the price of wool was tolerably steady. What estimate then may we form of the present value of a Merino flock, now that the price of imported wool is nearly treble its value in 1808,† and

G   no

---

* Vide page 33 of the Treatise.

† The following prices of imported Spanish wool are extracted from the New London Price Current, published by
Mr.

no immediate probability of its returning to its old level? The importance of the subject cannot be too strongly impressed on the mind of the public. At no former period of our history could the Merino breed of sheep be adopted in this country with so ample a prospect of remuneration as at the present moment. It will be in the recollection of the reader, that on Bonaparte's personal entrance into Spain, one of his first measures was to order the removal of 2,000,000 of sheep into France. Now the gross number possessed by the Spaniards, was never estimated at more than six millions,* which

---

Mr. John Arch, Capel Court, near the Royal Exchange, of the under-mentioned dates:—

July 1, 1808.

	s.	d.		s.	d.	
Leonesa wool . .	6	7	to	6	9	per lb.
Segovia	5	6	to	6	6	

June 16, 1809.

	s.	d.		s.	d.
Leonesa	16	0	to	20	0
Segovia	14	0	to	18	0

* Vide page 20 of this Treatise.

which, when we consider the havoc and devastation made by the French, in addition to the two millions sent out of the country, may therefore be considered without exaggeration as being reduced to half the original number. Even if the French could be expelled Spain to-morrow, the supply of wool that we should receive from thence, under the preceding circumstances, must, of necessity, for some years to come, be very inadequate to supply our demands.

Hitherto I have considered the subject solely as it respects the acquisition of a superior quality and increased quantity of wool. I shall now proceed to consider the carcase of the Merino and Anglo-Merino sheep, and I have great reason to think that the instances I shall produce of the amelioration already effected in this point, are such as will induce every candid and considerate person to believe, that as much improvement may be made in the carcase, by attentive breeders, as has already been atchieved as to wool. The following account of experiments made by Morris Birkbeck, Esq. of Wanborough, near Guildford, Surrey, will prove to the reader that the advantages of a fine fleece

are

are to be gained without any sacrifice of such properties, peculiar to our native breeds, as it is the wish of the English breeder to preserve. Mr. Birkbeck's experiments have been conducted on a large scale. He has, since the year 1803, annually crossed a South-Down flock of 6 or 700 ewes with *Merino-Ryeland rams* purchased from Dr. Parry, to whom he communicated in 1807 the result of his experiments as follows:

" The fleeces of the first cross (washed) are to the parent South-Downs, as 6 to 5 in weight, and as 3 to 2 in value per pound.

Thus, 100 South-Down fleeces, 2½lbs. each, at 2*s.* . 25*l.*
    100 First cross ——— 3 lbs. —— at 3*s.* . 45*l.*

", So much for wool, and were it not for the air of extravagance it might give my statement, I should add, that there is an evident improvement as to usefulness of form, and disposition to fatten, in a large proportion of individuals. I had the courage to exhibit at Lord Somerville's shew, in March last,* five ewe-hogs from your

rams,

---

* This letter was written in 1807.

rams, and the honour to bear away the prize from all competitors, by the merit of carcase and fleece jointly. On the whole, I believe that *the improvement of the wool may go on, without detriment to the carcase*, until we shall obtain a breed of sheep with Spanish fleeces and English constitutions, but I am also convinced that this must be the result of careful and judicious selection."*

Now it appears from this report, that in the wool produce of 100 South-Down sheep, *from one cross only* with Merino Ryeland rams, there was £20 increase of value, which is equal to £80 per cent. profit; and that too on a carcase, which, *instead of being deteriorated*, was for the most part improved by the admixture.† It is highly gratifying to observe in this instance the great increase of value of wool from the Anglo-Merino ram, without the intervention of either

the

---

* Communications to the Board of Agriculture, vol. 5, p. 538.

† Ibid

the pure Merino ram or ewe, and it confirms
(though no confirmation can be wanting to those
acquainted with his character) all that Dr.
Parry has written to prove that equal, or even
superior wool, is to be obtained on Anglo-Merino
sheep, *with an improved shape* and carcase, than
on the pure Merino; and that, when obtained,
it is no more necessary to refer to the pure Merino with a bad shape, to keep up the improved
state of the wool, than it is for the Newmarket
breeders, who, by the Arabian cross, have acquired such horses as Mask, Eclipse, Highflyer,
&c. &c. improved in size, bone, wind, and
every desirable quality by the crossing already
made, to return to the Arabian, Barb, or Turk-
ish stallion.

At Lord Somerville's shew in March last, Mr.
Birkbeck produced some very fine wethers (for
which he obtained one of his Lordship's premiums) which, in point of shape and condition,
quite astonished every one. They gave general
satisfaction, and also weakened so much the prejudices of some gentlemen present, who have
to my knowledge been many years deeply en-
gaged

gaged in the Dishley or new Leicester breed,
that they assured me they would immediately
try what they could do with the Merino. The
particulars of these wethers will be found in Mr.
Young's annals. Mr. Birkbeck, in answer to
a letter I troubled him with a short time since,
states as follows: " As far as the experiment has
proceeded under my management, the introduction of the Merino fleece appears to be com-
patible *with all the qualities* that we have been
accustomed to esteem in our English flocks."

I have heard that very fine wool has been produced, and some tolerable carcases obtained,
from a cross with the Wiltshire breed; and I
incline to that opinion, so far as I can form any
judgment from the specimen of a wether, belonging to his Majesty, shewn at the last sale at
Kew, and at Lord Somerville's spring shew, and
from a few at my neighbour's, the Rev. James
Willis, of Sopley, which, though descended
from a small ram of mine, are very large, and
will be when fat, at least 20lbs. per quarter. If
any gentlemen, who have tried the above cross,
or any other, find advantages in them superior
to what is stated to arise from the Ryeland or
South

South-Down, I hope they will favour the public
with an account of their experiments. It appears by a letter in the Agricultural Magazine
for March last, that Mr. Hose, a well known
Leicestershire breeder, has had the courage to
make the experiment in that county of crossing
the Merino with the Leicester; and that he has
declared his Merino-Dishley wethers fattened
more kindly than his pure Dishleys of the same
age in the same pasture, and that they produced
as much wool as the Dishley, and of double
value.

Since the introduction of the Merino breed
into this country, the primary object has been
to multiply them, and therefore the carcase has
hitherto been only a secondary consideration.
It has however already been remarked in the
introductory chapter, that Lord Somerville has
much improved the shape of his flock. I have
seen several of his rams as well grown as might
even satisfy Dishley breeders.* And if the
throatiness,

* In a new and enlarged edition of Lord Somerville's
" Facts and Observations on Sheep, Wool, &c." (published
while

throatiness, or dewlap, which is often found in these sheep, and which is merely a duplicature of the skin, be a fault, Lord Porchester, it is said, has entirely got rid of it. Dr. Parry informs us, that though he has bred indiscriminately from all his ewes, and universally preferred those rams which had the finest fleeces, his sheep are in general much shorter in the legs and neck, rounder in the barrel, and wider in the loin, than is usual with the pure Merino.* The ewes appeared to me when I saw them about three years since, to have very much the symmetry

---

while this sheet was passing through the press) his Lordship has favoured the public with an elegant engraving of his celebrated ram, whose symmetry of shape cannot fail to excite the admiration of every beholder. I avail myself of this opportunity to apprise the reader that the references which are made in this Treatise to his Lordship's "Facts and Observations," apply to the first edition of that work, published in 1803. It is necessary to bear this in recollection, as the pages of the two editions do not answer, in consequence of the enlargement of the present.

* Communications to Board of Agriculture, vol. 5, p. 467.

symmetry of a Ryeland flock, and from their velvet faces, denoting the superior quality of their wool, were beautiful.

In respect to shape, let us however consider more fully its importance; and this consideration will also lead us to a comparison of the merits of the Merino with the Dishley breed. "The capacity of quick growth and early fatness (it is observed by Dr. Parry) is generally supposed to be connected *with particular circumstances of form* and proportion. These circumstances Mr. Cline, in a paper addressed to this Board, has attempted to explain on phisiological, and, as it appears to me, just principles. He sets out with considering the external form of the animal as being of consequence, only as it is indicative of the internal structure. The power of converting food into flesh chiefly depends on the relative size of the lungs, which must be proportioned to the mean circumference of the chest of a given form, which should be as nearly as possible circular. This structure will necessarily influence the distance between the

the shoulder-blades and fore-thighs, producing a broad chest. The pelvis in the female should be wide and deep for the purposes of easy gestation and labour, and therefore for the production of sound and healthy offspring. On the breadth of the pelvis will depend that of the loin and the goodness of the hind quarter, which comprizes the size and distance of the hinder thighs. From these points must arise a straight back and large barrel, which regulate the dimensions of the stomach and intestines, and therefore the power of taking food. Large bones are not a sign of strength, but rather connected with a sort of ricketty disposition, which always implies weakness, and an incapacity of nutrition; besides which, they are not the food of man. The smaller, therefore, they are, the better. This principle strongly applies to the head, which, when large, endangers the ewe at the birth. Horned heads are much heavier than those which are hornless. Long legs are only an excess of offal, and require a proportionable length of neck, in order to permit the animal to feed. Muscles constitute what is commonly called lean, and are the in-

instruments

struments of motion. They should therefore be full and large."*

All these points are so far united in the Dishley sheep, that as a creature to fatten quickly he is unrivalled; but at the same time his early proneness to obesity unfits him for obtaining a large proportion of well-flavoured muscular substance. The fat on the loin in this breed being to the lean as 5, and sometimes as 6 to 1. For what then (asks Dr. Parry) is a perfect Leicester sheep fit? On rich land he is calculated at an early age to produce for eating that which cannot be eaten, but which is good for the manufacture of soap and candles.†

The Dishley sheep is naturally of an indolent disposition: any fence confines him: he moves for food to a short distance all around him, soon satiates himself, and then lies down to rest. His whole object, and the very end of his existence, is to fill his belly; and therefore,

from

---

* Communications to the Board of Agriculture, vol. v. p. 462.
† Ibid. p. 466.

from his habits and make, he requires rich, luxuriant herbage. Placed on poor or moderate keep, where a bellyfull is only to be obtained by exercise, these sheep starve, while, on the contrary, the Ryeland or South-Down will keep themselves in tolerable condition. The Dishley breed is therefore, comparatively, unfit for nearly half the pasture land in Great Britain.*

Here then is a striking instance of the superiority of the Merino breed and its admixtures over the Dishley, as the former will keep themselves in good condition on moderate pasture, and yield a very considerable profit by their fleeces. In this respect great advantages may be derived from keeping a dry flock.

Most of my neighbours who have Merino flocks as well as myself, and have been attentive only to the improvement of our wool, and increasing our stock to cross, have in general sold

off

---

* Communications to the Board of Agriculture, vol. v. p. 465.

off all our wether lambs. One gentleman,* a
neighbour and friend of mine, who keeps a dry
flock, has about 150 Anglo-Merino wethers, the
greatest part of which are, both *in point of shape
and carcase*, equal if not superior to any South-
Down flock I have ever seen; and the individuals
of this flock are chiefly of the second cross, or
three quarters Merino, and some of the fourth
cross, and of my breed. I have also myself about
150 wethers, all two-toothed or hogs, *i. e.* lambs
of last year, which are in general very handsome,
though still higher bred than the last mentioned
flock. Both flocks have this winter lived extremely hard: mine have certainly never fed
upon land worth on an average more than ten
shillings per acre, and have had very little hay
or turnips.

A striking peculiarity in favour of the Anglo-
Merinos is, that the wethers generally exceed
at maturity the size of either the Merino sire or
Ryeland dam. A Ryeland ewe seldom exceeds
15 lbs.

---

* The Rev. Dr. Wyndham, of Hinton, near Christchurch,
Hants.

15 lbs. per quarter, or the pure Merino ram 16 lbs.; but the majority of these wethers, bred upon tolerable land, will, I have no doubt, produce when fat from 17 to 20 lbs. per quarter, and several very much more. They have a great disposition to fatten, and will doubtless, when better known, be great favourites with the butchers, from their habit of producing much inside fat; which propensity, with their shape, I conceive they take in a great degree from the Ryeland dam, though the pure Merino ram, as far as I can find from those who have killed them, or I can judge from several specimens which I saw this year, which were killed by Lord Somerville for his public dinner, is not without great claims on this score. The kidneys of all these were well covered, and the quantity of rough fat very considerable. I was much gratified in seeing the carcases of these sheep, for while the pure Merinos remain at the high price they now bring, it is not probable many will be slaughtered for some years to come. Dr. Parry appears decidedly of opinion that the Anglo-Merinos take their shape and what relates to the carcase, chiefly from the ewe,

and

and the properties of the skin and fleece from the ram. In confirmation of my opinion as to their tendency to produce internal fat, I observe Mr. Tollet states, that a half-bred Merino wether, weighing 18¼ lbs. per quarter, produced 18¼ lbs. rough fat; and that while his best South-Down wether weighed 22¼ lbs. per quarter, and had 18 lbs. of rough fat, a Merino-Ryeland wether of the same age, of the first cross, fed with the former from a lamb, weighed 27 lbs. per quarter, and had 23 lbs. of rough fat.

It is a point which has been very much disputed, whether most profit is to be derived by the farmer and grazier on a small or a large animal of the same species. It appears to me that the weight of argument is much in favour of the former, as will be exemplified presently in respect to the Ryeland sheep; but as the discussion at large of such a subject is unconnected with my present object, I shall merely observe here, that the Merino-Ryeland wether comes to a weight always equal, and in general superior, to the pure South-Down equally kept. It is

mentioned

mentioned in the letter last alluded to, that a Merino-Ryeland wether of the first cross, three years old, bred by Mr. Tollet, and presented by him to Mr. Coke, of Norfolk, and killed at the last Holkham sheep-shearing, produced 33 lbs. per quarter.

From what has already been stated, it is evident that the Anglo-Merinos possess a disposition to fatten, equal to that of their pure English ancestors: and as it has been ascertained by experiment that our native South-Downs are as profitable to the farmer for general purposes as the Dishleys, it follows that in this respect the Anglo-Merino breed, of the South-Down cross, as much exceeds the Dishley, as it does the South-Down from which it is descended.

In respect to the comparative merits of the South-Down and Dishley breeds, it may be necessary to enter into the subject somewhat more fully. On some particular, strong, deep soils, the Dishleys may be the most profitable of the two; but the following result of experiments made in the West of England proves decidedly

H   the

the superiority of the South-Downs. These experiments, made by Mr. Billingsley and the late Marquis of Bath, are detailed in the Bath Society's Papers, vol. vii. p. 352, and vol. viii. p. 371. Mr. Billingsley's experiment was tried on six sorts of sheep, the relative superiority of which is thus classed by Mr. B.

First, South-Down, or Mendip,
Then Dorset,
Gloucester,
Leicester,
And lastly, Wilts.

These sheep were all two-toothed wethers, about a year and three quarters old, when sent to Mr. Billingsley in January, and killed the December following. After having given an exact account of the quantity of food consumed by each breed, Mr. Billingsley reports as follows:

" Uninfluenced and unbiassed, I waited with anxiety the result of an experiment which I considered as fraught with consequences of the

firs

first importance to the breeding counties of this kingdom, and if it has not been so conclusive as might have been wished, no blame is, I trust, imputable to me. I cannot agree in opinion with the gentlemen to whom the examination of my sheep experiment was committed. If I recollect right, they gave the preference to the South-Downs, and after them to the others in the following order, viz. Gloucester, Leicester, Mendip, Wilts, and Dorset. Now it appears to me, from the nett produce, and also from the quantity of food consumed, that either the South-Down or the Mendip should take the precedence, and that they should rank, as stated before, South-Down or Mendip first, Wilts last. The difference in the value of the skin and fat is not sufficient to alter this conclusion. At first view the Gloucester appear to produce most profit; but when it is considered they ate nearly one quarter more food than the Mendip, and one eighth more than the Dorset, such an inference would be erroneous. To the nett profit should be added 4 or 5s. per head for manure, as they were regularly folded. I think the long wool was over-rated in comparison with

H 2   the

the short. The result of the experiment was not so favourable to the Leicester as at the commencement I thought it would be. They were sent in high condition, and had from their appearance been extremely well kept. The change of food and climate appeared to affect them more than the other sorts; and though they were fed with hay of prime quality, and turnips perfectly sound and sweet, they invariably lost weight the first four months, nor did they in the subsequent summer months exhibit any great progressive improvement, as the statement plainly shews. One, indeed, was unhealthy, and had when killed a defect in his lungs, for which some allowance should be made.

" The Gloucester, or Cotswold sheep, appeared to be the offspring of a cross with the new Leicester, and consequently approaching very nearly to the same species, only in a larger frame. They consumed more food, grew more, and seemed to be a hardy useful sheep.

" The Wiltshire were a tall, bony, thin-car-cased

cased sheep, fit to walk two or three miles to a fold, and to be kept till three or four years old, for the purpose of manuring a Down farm. They ate ravenously, increased greatly in size and weight, but did not fatten.

" The Dorset, the South-Down, and the Mendip approach nearly to an equality in point of profit, and may be considered as valuable sorts both to the breeder and grazier; but were I to take my choice of a flock calculated to endure severity of climate and scantiness of pasture, I should prefer either the South-Down or the best sort of native Mendip; and in this idea I am justified by observations made in the course of this experiment. In the winter season, when the Leicester, the Cotswold, the Dorset, and the Wilts were unceasingly devouring hay and turnips, the South-Down and Mendip were traversing the field in search of the scanty pittance of grass then to be found; and I verily think that their wintering was not worth so much as the others by three or four shillings a head.

"J. BILLINGSLEY."

In the other experiment, made by the late Marquis of Bath, the most profitable breed of the six appears also to have been the South-Down, then the Mendip, Cotswold, Wilts, Leicester, and Dorset. On this subject Dr. Parry says, " From a manuscript paper given me by Mr. Davis,* who conducted this experiment, I have it in my power to add, that the *Ryeland sheep, which he also brought into competition, beat all the other sorts;* but that no mention was made of them in the report presented to the Bath Society, because they were not specified in the premium."✝

After this very satisfactory account of the Ryeland sheep, instead of being surprised at, or unwilling to believe the favourable accounts given of their Anglo-Merino descendants, it appears to me it would be very extraordinary they should prove otherwise than they have done.

---

* Agent to the late Marquis of Bath. A gentleman of the utmost respectability, and well known in the agricultural world.

✝ Dr. Parry's Facts and Observations, p. 44.

done. The following particulars will evince their hardiness and adaption to the purposes of folding.

Colonel Mitford, of Exbury-house, near Beaulieu, and Colonel Serle, of Chilworth-lodge, near Southampton, have for several years crossed the Ryeland ewes with Merino rams, to a very considerable extent, and have now both of them large and beautiful Anglo-Merino flocks, which have been constantly worked very hard in the fold with great success, though they have lived in general extremely hard, and been occasionally depastured upon the commons in the neighbourhood of the New Forest. To their hardiness and excellence I have been repeatedly an eye witness, and as I do not fold my own, I take the liberty of mentioning these flocks, which I have no doubt any gentleman is at liberty to see; and also that of Colonel Conynghame, of Maltshanger, near Basingstoke; the following extracts from whose letters will be found to corroborate what has been already advanced on the subject. The Colonel's first letter was written to me about the time of the last

meeting

meeting of the Christchurch Agricultural Society of which he is a member in answer to
some questions I had asked him, and is dated,

" *Maltshanger, near Basingstoke,*
2d Nov. 1808.

" I have not yet received the Merinos my
friend Sir Thomas Dyer was so good as to say
he could procure for me. In the present state
of Spain, and the difficulty of obtaining a pas-
sage for the sheep, it is no easy matter to get
them over.

" I never had an idea of changing my flock
of Ryeland-Merinos, as they continue to answer
my most sanguine expectations. I have now,
including the last fall of lambs, about 18 score
half-bred, or the first cross. My half-bred ewe
tegs are this season, for the first time, put to the
Merino rams, by which I hope to procure five
or six score three-quarters bred next spring. I
do not allow my ewes to be put to the ram till
they are a year and a half old, as I think it ruins
the animal to allow them to produce at an
earlier age, and the offspring frequently dies,

or

or is puny. Thus every cross takes me two years.

"I have, including my old Ryeland ewes, which I am now fatting off, and no longer breeding from, near 500 sheep, all upon turnips. I have sold all my wool, and at such prices as to leave no doubt of the profit to be derived from a Merino flock over *all others*.

		s.	d.
My Ryeland ewes' fleeces, at .		2	4 per. lb.*
Ryeland Merinos, first cross		3	3
Lambs' clippings . . .		1	6
And an odd lot, cut from Dorset ewes, South-Down wethers, and horned wethers . . .		1	3

"The Ryeland-Merinos weighed 3¾ lbs. each fleece, one with another. The whole was washed on the sheep's backs, with peculiar care.

"The

---

* These prices, it will be perceived by the date of the Colonel's letter, were some time before the late advance on wool.

" The whole of my flock, (those fatting excepted) *lambs* and all, go regularly to fold, winter and summer, upon the fallows. They stand their work as well as any South-Down flock whatever; and confessedly by all my neighbours no sheep live harder, or are more healthy. The loss I have experienced by casualties, is very trifling, and infinitely less than that generally the case in the neighbouring flocks, in proportion to numbers."

The second letter Colonel Conynghame favoured me with, is dated 13th May 1809, and states,

" I have the pleasure to acknowledge the receipt of your letter, and am happy to have it in my power to assure you that my Anglo-Merino sheep continue every way to improve, and thrive upon this soil. My management of them is exactly the same as I mentioned to you in my letter of November last, and they have stood all the bad weather of last winter and spring, better than any of the neighbouring flocks. I have now about 100 lambs of the second cross

(*i. e.*

(*i. e.* three-quarters bred) and they promise to carry wool of a very superior quality. Every step made with the Merino rams, shews that there is no doubt but that wool may be grown in this country, by several continued crosses of the Merino ram, equal to that generally imported from Spain. As yet mine is the only Merino flock in this part of the country, but several farmers begin to see the profit to be derived from the Merino cross. I have had many enquiries made this year for full blood, and well-bred rams to cross upon their South-Down ewes next Michaelmas; and I have no doubt but that in a very few years the Anglo-Merino breeds will supersede every other breed of sheep in this part of the country, a circumstance most desirable for the community at large, in my opinion, from the money which will be diffused through the country, instead of being sent out of it for fine wools. I am happy to learn you mean to communicate to the public, through the medium of the press, the valuable information you must have obtained respecting Merino sheep, from your extensive practice as a breeder, and I trust it will be appreciated as it ought to be."

Charles

Charles Jenkinson, Esq. M. P. of Beech
House, near Christchurch, a neighbour of mine,
has also a very valuable flock, and several of the
pure blood. In the beginning of April last, on
going with another gentleman to look at his
flock, which were finishing his turnips, we were
astonished to find amongst his tegs (or hogs as
they are often called) a number of horned sheep,
looking miserably ragged and starved. On
sending for his bailiff, we found that having
more turnips this season than his master's flock
could consume, he had taken in 40 ewe-hogs of
the Dorsetshire breed, to keep, from Michaelmas
to Lady-day, at 8*s.* per head, or £8. per score,
the usual price this last year. On our returning
again to the fold with the bailiff, and my stating
that I had never seen such a set of wretched ob-
jects as they were, compared with his master's
own hogs, and that I would not have them to
pay for their keep, he assured me (and he is a
man of character) that they had fared in every
respect the same as Mr. Jenkinson's Merino and
Anglo-Merino tegs; and had run with them from
the time they came to his farm at Michaelmas.
There were two Leicestershire hogs also in the
same fold, looking but little better than the

Dorsets,

Dorsets; while on minute examination, we found the whole of the Merinos and Anglo-Merinos in good health, and not one broken fleece amongst them.

I will not attempt to give stronger or more instances of the health and hardiness of Anglo-Merinos than the foregoing; nor can I give a fairer challenge to investigation than by naming the flocks of the four gentlemen I have mentioned, all of whose seats and farms are on high and exposed situations, beautiful and delightful for every other purpose, but that of folding sheep in winter; and the land about them is in general such as has never, till within these few years, been used for sheep, or thought capable of carrying them.

When the fleeces of the Merino, and high-bred Anglo-Merinos are examined, it will be found from the fine and close texture of the wool, that there is a wonderful capacity of resisting the bad effects of the most severe weather. The extreme exudation from the body yields an oily, yolky moisture, at the interior of the

the fleece, which, as it approaches the outside, mixes with soil and dirt, and forms a sort of coat of mail equally impregnable by rain, or cold winds. I trust, after what has been shewn, we shall never hear more of the delicate constitutions and the tenderness of this sort of sheep. It remains now to consider them in another, and most important, respect.

Every person well acquainted with Merino sheep, speaks of the excellence of the mutton; and I have never yet met with any individuals who have *really* tasted it, or that of any of the Anglo-Merinos, who are not of the same opinion. Yet that persons foiled and defeated in every other attack made upon them, may be base and weak enough to make some attempts to decry them in this respect, it is not unnatural to suppose; but truth and justice, though slow, are sure at length to prevail, and the effect will ultimately be, as it has been in other instances, that by raising the curiosity of the public, and inducing persons who would not otherwise have thought about the matter, to make

the

the experiment of tasting it, such opposition will certainly defeat its own purpose, and bring the mutton into the estimation it deserves. How any idea should get credit that this mutton was not good, or why it should be otherwise, I cannot imagine. We have hitherto always reckoned, and found, that the finest and shortest woolled English sheep have produced the best mutton. The inference might therefore naturally be expected to correspond, as it does, with the fact, that the finest woolled sheep of all would produce the finest and best mutton, which is so much the case in respect to Merinos, that I have known it preferred to venison.

The testimony of Sir Joseph Banks, as to the excellence of the Merino mutton of the King's flock, has already been adduced in the preceding chapter. Lord Somerville observes, " Whether as an article of food for those who are robust, or those who are delicate; even at the early age of 18 months, when mutton is usually thought indifferent, it is nutritious and exquisite in flavour. There is a firmness in the spine fat; a richness and deep colour in the gravy;
and

and a fine texture and tenderness in the grain, which must command customers, and ensure to this breed the good will of butchers, wherever they may be situated."*

Mr. Sheppard, in his letter to Sir John Sinclair, in December 1806, concludes as follows, " I last week sold half a score six-toothed wethers, of the first cross, to the butcher, fatted on grass and hay, for £22. 15s. exclusive of their wool. I cut 5¼ lbs. of wool in the grease, from one sheep, which when clean scoured, produced 3¼ lbs. well worth 5s. per lb. The other fleeces averaged rather less. This great growth since the last shear time, I attribute to the fatness and size of the sheep, which weighed 21 lbs. per quarter. The rest averaged 19 lbs. per quarter. The whole value of carcase and wool exceeding 60s. exclusive of 15s. the value of the last fleece at shear time. *I readily obtained a penny a pound more than the market price, on account of the beauty of the meat, and its great*

---

* Lord Somerville's Facts, *p.* 28.

great fatness; and I must not omit the testimony of both amateurs and adversaries to the mildness and excellency of the mutton."*

One butcher in Bond-street has thought proper to find fault with this meat, and it is reported, has said it is no better than carrion, and inferior to other mutton 3*d.* per lb. If any of the foregoing or following assertions of some of the most respectable characters in England, are entitled to any credit, this man must have been very unfortunate in what he has killed, or he must have invented the story with some very malignant intent. On this subject, Lord Somerville, at his last public dinner, according to the reports given of his speech in the daily papers, expressed himself as follows:

" General conclusions are never to be drawn from individual cases, and admitting the fact that one or two sheep might not have turned out well, this was cause for further inquiries, but not for

I   universal

---

* Communications to the Board of Agriculture, *vol.* 6, *p.* 73.

universal condemnation. An old ram may be very excellent mutton; but I have yet to learn what the breed is which will produce it. It would be the excess of injustice to cry down one of the best breeds of sheep which we know on the surface of the earth, the South-Down, merely because they turn out sometimes so yellow as to be unsaleable, except at reduced prices. It is well known that all the wethers which I have fed off for several years past, to the amount of some hundreds, with the exception of a few sent to London, have been retailed at Taunton market on my account, and on the average have produced one penny per pound above the prices of other mutton. That large as the supply was on each day, it was bought up with an avidity that was remarkable, was generally bespoke on the previous market-day, and was actually purchased on commission, and sent off by coaches to the adjacent counties, an instance of preference that perhaps has rarely occurred before. This continued to be the practice until the day my stock quitted the county. Some few have been sold in this city, by Mr. Henry King, of Newgate-Market, whose books

will

will prove what the demand for it is; and Messrs.
Thomas Leybourne, Garment, &c. confirm
that it is held in the highest estimation. The
supply for this annual dinner was five sheep,
but, for the two last seasons, when the mutton
of the pure and unmixed Merino breed only
was served up, the demand for it has been
such as to cause a request that the number
of sheep might be increased to eight in place
of five. The Duke of Bedford, with that
liberality which marks every act of his life,
knowing the effect produced by these misrepre-
sentations, being in the breed of South-Downs,
has (unsolicited by me) sent, together with a
Merino sheep, which breed he keeps solely for
the use of his own table, a note which I beg to
read—

"*Woburn Abbey, Feb. 28.*

" MY DEAR LORD,

" I HAVE to ask of you to excuse the
liberty I have taken in sending you a wether of
pure Merino blood, three years old, weight of
the carcase 43 lbs. I am aware that there is no

I 2   small

small prejudice against this mutton, and from no less authority than an eminent butcher in Bond-street, but from the little experience I have had from the wethers I have killed here, I can pronounce the mutton to have proved uniformly good, and I have not the smallest doubt, that if you should think fit to feed any mutton epicures with any part of this sheep, *calling it Welch, or Scotch,* that it will be called excellent mutton.

" BEDFORD.

" P. S. The sheep was killed yesterday."

" *The Rt. Hon. Lord Somerville.*"

Mr. Thompson, of Red-Hill Lodge, near Nottingham, remarks that " those who continue to rail against this mutton, have probably never tasted it; for many persons, originally prepossessed against it, have acknowledged their opinions to be groundless on trial. At Bath (says he) it is thought to approach the flavour of venison more nearly than any other kind of mutton, and those, who lately tasted the wether slaughtered on my shearing-day, unanimously allowed the

the quality to be excellent." When the late Marquis of Exeter died, about 300 sheep, possessing various degrees of Merino blood, were sold. Several of these fell at a cheap rate into the hands of Mr. Pollard, a butcher, at Stamford, as few persons except Leicester and Lincolnshire breeders attended the sale. " The work of slaughter now began (continues Mr. Thompson) and no sooner had the epicures of Stamford tasted this mutton and lamb, than the destruction of every animal with Merino blood in his veins, to be found in the neighbourhood, was resolved upon. Pollard said they proved remarkably well, and that the kidney was all fat."*

From the aptitude of the Spanish race to fatten, in an equal degree with any of our native breeds, the admirers of fat men and fat mutton may console themselves, if the Merino breed should prevail, that they may procure as large, as fat, and, to say the least of it, as well-flavoured

---

* Letter to the Marquis of Titchfield, 1808.

flavoured meat, from the descendants of this breed, as the fine Leicestershire herbage has yet produced from any breed whatever, instead of deriving " *only a small quantity of lean* Spanish mutton, in return for *the immense quantity* of herbage these sheep would consume;" so that the alarmists will be in no danger of being called upon to " resign their fat mutton, *and their own fat into the bargain,* and all for the sake of covering their lean sides, with a fine coat made of Spanish wool,"* as has been suggested. On the contrary, mutton epicures are sure to benefit by the introduction of these sheep, for though they have all the disposition to ripen, and come to the knife as soon as any other breed, the fine wooled wethers are likely to be kept till they are five or six years old, as their wool will abundantly pay for their keep. I have killed a great many Anglo-Merinos myself, all of which have turned out to my most complete satisfaction, and though I have frequently produced

the

---

* Vide Agricultural Magazine.

the mutton at my table to gentlemen by no means friendly to the breed, having imbibed all the prejudices usual against innovation, I have never met with one who did not allow that the mutton was excellent.

CHAP

CHAP. IV.

Containing some further arguments in favour of the introduction of Merino sheep; with practical remarks on the crossing and management of the Anglo-Merinos; on the proper time of putting ewes to the ram; the proper rams to be used; on the propriety of shearing lambs, and of housing sheep in bad weather, and at other times; and on the effect food is supposed to have upon wool; shewing that no food, properly used, will affect it in any material degree.

HAVING fully, and I trust satisfactorily, answered all the objections hitherto urged against the introduction of Spanish sheep, and the adoption of the Anglo-Merinos as a new breed; and having proved that instead of their possessing tender constitutions, unsuited to our climate,

they

they are strong and healthy, capable of endur-
ing all climates, seasons, and every sort of
treatment to which the hardiest of the English
breeds has ever been subjected; that the wool,
instead of degenerating, is improved on English
soil; and, under the care of English breeders,
produced as fine and as fit for every purpose of
manufacture as the finest piles from Spain; that
the carcase, in point of shape, is improved, and
still improving; and that, by the Ryeland and
other improved crosses, a disposition to fatten
and to come to early maturity is manifested,
equal to our most admired breeds; and that
their mutton is exquisite; it will now, I flatter
myself, be admitted that the august promoter
of their introduction and distribution, and the
noblemen and gentlemen who with such zeal,
industry, and ability, have seconded his patriotic
views, and whose names I have before recorded,
have not been led away, as farmers and others
have been industriously taught to believe, 'by
enthusiastic notions, from which no good can
result,' but have acted with the soundest judgment,

ment, and most prudential foresight, caution, and consideration. And great must their gratification be, on finding what now incontestibly appears to be the fact, that the subject to which patriotism alone first directed their attention, is likely not only to exceed their most sanguine expectations, but to prove a mine of wealth to the kingdom at large, and add to the comfort and advantage of every individual in it. The importance of the object is, I believe, now very generally admitted; but if there are any who still doubt, or have not attended to the matter, let them consider, as Mr. Thompson very properly observes, that "the French government is paying close attention to the breed of Merino sheep, and is warmly encouraging superfine manufactures through every department of the vast territory under its controul. Even when a peace shall take place, we may reasonably suppose that France and her dependants, now consisting of nearly a hundred million people, will have a preference on the part of Spain, and other continental powers; and why may not this immense population, under the able guidance of a

government

government bent on our humiliation, consume all the superfine produce of the continent? Not a yard of superfine cloth can be made but out of Spanish wool only, or of that produced by crossed breeds, brought to the same perfection by judicious perseverance. Nothing would, therefore, in the case just mentioned, remain for England but to obtain the precious article here and there, by outbidding her rivals in price. The consequence would be, that this extra price of the raw material would fall upon the manufactured commodity, and forbid the possibility of British goods finding a profitable market, or being able to cope with those of France and her connexions. No such case could occur if we provided ourselves with Spanish wool by growing it within our own shores. The period is therefore probably at hand, when it will depend upon British farmers whether the British manufactures shall continue to maintain their proud pre-eminence, or shall dwindle into contemptible insignificance, bringing consequent distress upon thousands now employed in them. The individual advantage of the farmer, and the

prosperity

prosperity of the nation at large, are equally involved in the measure."*

Whatever may be the issue of the present convulsions in Spain, and even should they terminate favourably for this country, surely there can be no reason why we should not make ourselves, as far as possible, independent as a nation, for an article which employs so large a portion of our population, and forms so material a branch of our commerce. In calculating the advantages to the farmer, resulting from the Merino and Anglo-Merino sheep, I do not consider any fancy prices gentlemen may chuse to put upon these breeds, while they are scarce and difficult to obtain. I only consider the mere mutton and wool as such, and contend, that reckoning the latter at a very moderate price, and the mutton at the usual rate, the advantage is infinitely in their favour above every other breed. Having already given so many
instances

---

* Letter to the Marquis of Titchfield, President of the Newark Agricultural Society, 1808, pp. 23, 24.

instances in proof of this observation, I shall only add the following account of a three-shear Merino-Ryeland wether, of the first cross, bred and killed by Mr. Thompson:

		£.	s.	d.
His 1st and 2d fleeces, together 10 lbs. sold at 3s.		1	10	0
3d fleece, June 13, 1808, 7 lbs. 2 oz. at do.		1	1	4
Carcase . . . . . 87 lbs. at 7d.		2	10	9
Rough fat . . . . 18 lbs. at 6d.		0	9	0
		£. 5	11	1

This is by no means an exaggerated account, for both mutton and fat are below the market prices of the same article in this county, and the wool is rated much lower than a great deal of the same sort has been sold for, before the late extraordinary rise upon fine wools. Besides specimens of cloth, cassimere, and Norwich stuffs, made from Merino and Merino-Ryeland wool; and stockings and yarn from Merinos and Merino-Dishleys, Mr. Thompson produced at the Newark Agricultural Meeting for 1808, two bats made entirely from the wool of a Merino

rino-Ryeland ewe, which died during the lambing season, on which he observes, "I do not
exhibit these as proving the best purpose to
which such wool can be applied, but simply as
shewing one among many uses which may be
resorted to for the consumption of casualty
fleeces. Each of these hats contains nine
ounces of wool; and the manufacturing ex-
pences, with binding, band, and buckle, are
3s. 8d. Supposing the hatter, therefore, to pay
5s. per lb. for this Merino-Ryeland wool, when
scoured, the prime cost of each hat will be
6s. 6d."* It appears, too, that Anglo-Merino
skins take the morocco dye extremely well.

Mr. Thompson advises crossing the Sher-
wood Forest sheep with the Merino, and addu-
ces an instance wherein, by one cross only with
the Spaniard, the forester's fleece was increased
1½ lb. by the experiment. Thus, the forest ewe
yielded 2½ lbs. of wool, and the Merino-forester 4 lbs. Both fleeces were exactly of the
same

---

* Letter to the Marquis of Titchfield, p. 28.

same growth. The forester's was worth 5s. and the Merino-forester's sold for 12s.*

I strongly recommend to any farmer inclined to cross his flock and improve his wool, to begin at once with a pure Merino ram, or at least with one of the fourth cross, which is allowed to be equal to the pure blood. Full-bred rams are to be purchased or hired at much less prices than have been for many years, and are still given for the New Leicester. The average price of them at his Majesty's last sale, amongst which were many very excellent ones, was under £35, and they may be hired of eminent breeders, for the season, at a cheaper rate. As a ram is certainly fit for eighty ewes at least, even if he be purchased for £35, the cost of the lambs by him will not amount to more than 8s. each, which will be more than repaid by the increased value of their fleeces the first year. The ram will most probably serve many seasons. I have never known an instance wherein a person afraid

of

* Letter to the Marquis of Titchfield, p. 19.

of the expence, has begun with a half or three-quarter bred ram, because he was cheap, who did not afterwards regret that he did not commence with the pure blood.

It is supposed by Dr. Parry and other gentlemen, for whose opinions I have the utmost respect, that in crossing the offspring takes its shape, and the chief properties of the carcase, from the ewe; and those of the skin and fleece from the ram. If this be the case, little more is to be sought for in a ram than a fine fleece; but as I do not entirely agree with those gentlemen in theory (because my practice, as far as it goes, does not confirm it) I should recommend where it can be done with convenience, to attend a little to the shape and carcase of the ram. We look for the best shaped males, of all other animals, to breed from; and that we shall, in a few years, from the attention lately paid, and likely to be paid to these, have as well-shaped Anglo-Merino rams as those of any other breed, I have not the smallest doubt. I feel disposed to agree with Lord Somerville, that the bad shape of the Merinos arises in a great degree from the

mode

mode of treatment in Spain; from the long journeys they are obliged to undergo, without giving nature time to perform all its functions; and that (to use his Lordship's expression) they are absolutely 'hunted into deformity.' And this opinion seems to derive confirmation from the appearance of several of my pure Merino tegs and lambs, which, from being well kept and attended to, are certainly infinitely better shaped than either the sire or dam. I incline to think the observation of an old Warwickshire farmer, who (being pressed to alter his breed of sheep and adopt those of Leicestershire) declared that "all shape went in at the mouth," entitled to more consideration than the words appear to convey; for though I do not mean to assert that good keep will remove a dewlap, or diminish a head, I am sure that any young animal, uniformly well kept, will acquire a rounder, and in every respect better form, than one that is half starved or much hurried about; and that when a good form is once acquired, it is easily continued and propagated.

The time of putting the ram to ewes must

K   depen

depend entirely upon the climate in which they are situate, and the treatment intended for the lambs. My practice has been to put them to the ewes at the latter end of July, by which means I get my lambs about Christmas, which I think by far the best season. About this time I put all my ewes into small inclosed places, and supply all those likely to yean with sheds, (not close warm houses, but) such as will keep them dry and sheltered from cold cutting winds. I have shewn in the preceding chapter that this breed of sheep requires less shelter, from the closeness of their fleeces, than any other; but I think all sheep ought to be so managed, or at least ought to be less exposed than they generally are; and from want of attention to this, it appears to me (and which I shall treat of by and by) that most, or at least the most material diseases to which sheep are subject, originate.

But to return to the lambs. Amongst my reasons for having them early a material one is, that the weather is certainly not more severe about Christmas, than it usually is two or three months later;

later; that turnips, cabbages, and food of that description, are in a more succulent state, and produce abundance of milk; and when green food becomes scarce, and the ewes can get but little, and of course give less milk, at the latter end of March and in the month of April, the lambs are almost as well able to maintain themselves by eating hay as the ewes. Besides, my lambs are all fit to wean by the end of May, by which means I get all the ewes I mean to dispose of, or to kill, in good condition long before Michaelmas. I also get, by their falling so early, a tolerable quantity of wool from them. My lambs' fleeces last year averaged upwards of 2¼ lbs. and many of them produced 3 lbs.

The propriety of shearing lambs must depend entirely upon the time of their being yeaned. I am quite sure that my lambs improve much immediately after shearing, not only on account of being cooler and more comfortable all the summer months, but, by getting rid of the *hippobosca*, or sheep tick, they feed and rest much more undisturbed. And as I

K 2   always

always shear them about the end of July, if the weather suits, they get sufficient fleeces to keep them warm before the cold weather comes on, and I am persuaded they produce-nearly or quite as much wool at the following shear-day as if they had not been shorn, and that the wool of the second fleece is more regular and better.

Having occasionally taken lambs from some of my best bred tegs, from an eager desire to increase my flock, and these lambs coming much later than those from my ewes (for instance, some of them in April and even in May, which are called cuckoo lambs) I have not ventured to shear any of them; and I do not think that they in general produce so much wool as many of my shorn tegs, though they are not much, if at all, inferior in size. I contrive to keep these tegs and lambs as well as possible, and to wean the lambs early, and I do not find that I have at all injured my tegs by this practice; but I mean to discontinue it, as it is very inconvenient in large flocks to have lambs coming in at so many different periods, and to keep

them

them separate till they are able to make their way well with the stronger flock.

Castration, it is well known, has an effect on the horns of all animals, and in general gives them a feminine appearance. If adopted early, I apprehend it entirely stops the growth of the horn in the Anglo-Merinos, not one in twenty of my wethers having the least appearance of any. My practice is to draw the testicle, if the weather is tolerable, as soon as the lambs are ten or fourteen days old. Some, from accident, or from an idea of keeping them entire, have been permitted to remain till the horns have budded, or I am inclined to think nearly the whole would have been polled; and I am the more inclined to this opinion from this day examining another neighbouring gentleman's flock, who does not castrate till his lambs are three or four months old, when they are seared with a hot iron. Though precisely of the same breed with my own, from the Negrete ram, many of his wethers have large horns, and a ram-like appearance. I have never lost one by my method, which appears to me not only less

cruel,

cruel, but more safe and advantageous in every respect. I know there is an old prejudice in favour of the other practice amongst the breeders of polled sheep, that letting them continue rams a certain time, gives them a better loin, and a broader and fuller appearance. But were there any ground for such an opinion, which I do not admit, I think it is a very bad method, and particularly where horns probably attend its adoption. I am told it is practised in Dorsetshire, and among other horned breeds, to give what are called good heads.

With respect to housing sheep, I have always thought and found that warm and dry lodging is as material to all animals as food; or at least, that the best food, without it, is deprived of half its effect. Animals unconfined, or in a state of nature, always find shelter under banks, trees, or uneven ground, from cutting winds or driving storms; but confined in open yards, as I have often seen the cattle in this country, exposed to every blast and storm, or sheep penned in folds upon bleak hills, in long dark nights, for the sake of manuring

some

some few acres of hungry land, out of the reach of the dung cart, cannot possibly thrive, though fed with the best food, (which under such managers is not often the case) and the foundation is thereby laid for all sorts of diseases, of which they drop off at no very remote period, to the astonishment of their unthinking and merciless owner, who loses more in his flock than the land he folded will ever pay him. In Herefordshire, and some other counties, but more particularly the former, from whence we derive our finest woolled breed, the Ryeland, the farmers generally house their sheep at night. Extremes of heat and cold are alike injurious to sheep, and it is equally advantageous to provide them with shade and water in very hot weather, as with shelter in the most severe. There is no difficulty in housing sheep upon a small scale; and, in my opinion, more advantages than inconveniences will attend it upon a large one, when the matter is well considered, and a proper mode adopted. I am not sure that either heat or cold have any effect upon the quality of the fleece; but I am quite certain that keeping the

sheep

sheep in a healthy store state, has an effect both upon the quantity and quality; and Mr. Sheppard truly observes, that the wool of a half starved sheep, though it may be very fine, is sickly, and void of proof in manufacture.*

The first flock I saw housed, or I may rather call it sheltered, upon a large scale, was that of Colonel Serle, of Chilworth Lodge, to which my attention was directed about three years since, by the very obliging attentions of his friend George Compton, Esq. who, during the Colonel's absence with his regiment, has the management of his flock. He adopts it only at, and previous to the lambing season, till the lambs are strong, and in snow, or extreme wet weather. It is a common fold yard, round the barn door, surrounded by the same sort of sheds as are usually made for cattle, except that they are much lower; and it is divided by hurdles at the lambing season, in such man-

ner,

---

* Communications to the Board of Agriculture, *Vol.* vi. *p.* 71.

ner, for ewes in different states, as the shepherd thinks proper. The whole is littered with fern, or any sort of litter fit for no other use, and repeated as often as it is trodden properly, or become wet and stained. The quantity of valuable manure alone made by this plan, is infinitely more than sufficient to pay for all the trouble which attends it, exclusive of the great advantage to the health of the ewes and lambs, and to the increase of wool.

On this subject Dr. Parry says, " The most effectual method seems to be that practised in Herefordshire, of housing the sheep at night, in buildings which they call cots, and which, according to Mr. Marshall, are generally 'the ground-floor of a large building, which is chambered at five or six feet high. The size is, of course, in proportion to that of the flock. A yard square to a sheep may be taken as the medium allowance of room. Racks are fixed up against the walls; and in the larger cots, some of which will cot two hundred sheep, other racks are suspended across the middle of the room, and hoisted as the dung and

and litter rise. Their food in the cot is sometimes hay, and sometimes barley straw, but
most commonly *pease-halm*, a food which it
seems is particularly affected by sheep; a fact,
which the rest of the kingdom does not seem to
be fully possessed of. The halm, however, is not
always thrashed clean; the under-ripe pods being
frequently left unbroken for the sheep. The offal
is strewed about as litter; and the cot cleared
out once or twice a year; or as often as necessity, or conveniency requires.' " Dr. Parry observes also, from the account of the same writer,
and other good authorities, that the store-sheep
are cotted at night by many farmers, both in
summer and winter, and by others in the winter only. The lambs are always cotted, and
with great benefit; the fatting sheep never.
" Lord Somerville proposes, (the Doctor continues) instead of fixed buildings, moveable
folds, the hurdles of which should be thatched
next the wind, and others of the same kind
placed above, on poles, by way of pent-
houses."* This certainly appears an easy and
useful

---

* Dr. Parry's Facts and Observations, *pp.* 81, 82.

useful experiment. Now, for instance, if a temporary fold of this sort was run up in some dry, sheltered spot, in or near large turnip fields, or tracts of ground where the sheep were principally to feed during the winter; and straw or fern ricks, or any other kind of litter, provided during the summer and autumn; the immense quantity of manure the sheep would make upon the spot, by being housed at night, would alone, as I have before stated, amply compensate for the trouble; the whole benefit of which, the farmer otherwise loses in ditches, and under hedges, and close places, where sheep naturally creep for shelter. Sheep, in general well fed, are indifferent to cold; it is wet alone which they dread, and which injures them; and in treating on this subject, I do not allude to Merino and Anglo-Merino sheep exclusively, for I have, I trust, very sufficiently shewn, that from the very nature and oily condition of their fleeces, they are better proof against cold winds and rain than any breed whatever.

I have seen in Herefordshire, and some parts

of

of Berkshire, moveable cots or sheds set in the middle of a fold in turnip fields, which, where no other mode is adopted, I think highly commendable. They are formed by filling a common dung cart with straw, and thatching it. A sort of gearing used on waggons for carrying hay, in which staples are driven, surrounds the cart; and on it light withy hurdles, called in this country gate hurdles, are thatched, and hung within about two feet and a half of the ground, and extended a little, by poles, from the cart, to shoot off the wet. By putting a horse in the shafts, these moveable hovels are readily placed as may be most convenient. Were the principle once adopted, and the propriety of paying this sort of attention to sheep admitted, which I have great expectation it will be at no very distant period, I have no doubt the good sense and ingenuity of my countrymen would contrive many modes yet unknown to attain the object. Lord Somerville remarks, that foul, close sheep cots are injurious; that a free circulation of air is always found beneficial; and that the Merino sheep suffer more from heat than

than from cold.* Adopting the idea from the sheds prepared last winter for His. Majesty's Paular flock (which were long sheds from the walls of Kew Gardens, sloped to within about four feet of the ground, and fronted at the distance of about ten or twelve feet, with straw or reed walls, about six feet high, so as to leave a yard in front of the shed, and at the same time prevent any driving storms from entering) I built some furze ricks in the front of some rough sheds I run up against the banks of an old gravel pit, at a very cheap rate, which completely answered the purpose. Indeed a better fold yard I do not desire. I mention this merely to shew that shelter, only, is required, and that expensive buildings are by no means necessary to carry this plan into effect.

On the subject of sheltering sheep, Dr. Parry remarks, " Whatever effect such measures may have on the wool, there can be no doubt that they must be highly conducive to the preservation of the health of the sheep and lambs, more

---

* Lord Somerville's Facts and Observations, p. 42.

more especially in mountainous countries, where there is little opportunity of driving the flocks, during the severity of the winter, to warm and well-sheltered vallies. It must therefore be a great object to every proprietor of large fine-woolled flocks, to provide good accommodations of this kind, which will amply repay him for the expence of erection."*

The plan of housing, or sheltering, is not only attended with the good effects stated, but saves the lives of many individuals at the lambing season. It also secures sheep from being worried or hunted at night by dogs; or the lambs from being destroyed, as is often charged by foxes, (though I believe dogs are, nine times out of ten, the culprits;) or being lost, from being yeaned in inclement weather, in exposed situations. Hundreds of lambs, and many of the ewes, die every year, either from want of assistance at the time of yeaning, or cold taken after it. When sheltered, the shepherd

---

* Dr. Parry's Facts and Observations, p. 82.

herd can find them at all hours, to afford them assistance and protection. Besides, he will not be so unwilling to go to them, or to remain with them when necessary.

As a proof of these assertions, the year before last I permitted a neighbouring farmer, who has a large flock; to send four of his best woolled South-Down ewes to one of my Merino rams. On enquiring how he liked his lambs in the spring, he told me he had had " desperate bad luck," and had only one remaining, though they all came very fine. A fox had taken one; another died from the ewe being young and poor, and having but little milk; two, which were twins, were dropped in the night when the wind was high, and it froze very hard, and he found them dead and frozen to the ground, under a hedge, and he was afraid the mother would not recover. Another neighbouring farmer, who has a South-Down flock, on my observing to him that, notwithstanding the severity of the late winter, his sheep looked better than almost any I had seen, replied, " Sir, they certainly do. I have been this year a little in your way.

I let

I let my sheep, this winter, lie at the barn door, from turnips, and it is astonishing how much better they have done than ever I knew them. They eat a great deal of the wheat straw, and, by lying warm, and picking the straw and small ears, which they routed for latterly like pigs, did so well that I gave them no hay, and I shall certainly in future adopt some plan of this sort."

Many sorts of land, too, will answer a much better purpose to the farmer, by drawing the turnips, &c. to folds of this sort, littered, and carrying the manure so made to the land, for no better compost is made than that to which sheep contribute. Sheep attended to in this manner are also certainly less ravenous, and consume less food, with more good effect, than those exposed and compelled to perpetual action to keep themselves alive. I have lately only seen a book called the "Shepherd's Guide," published in 1807, on the diseases of sheep, their causes, and the best means of protecting them, by James Hogg, the Ettrick Shepherd, from which I

am

am induced to quote the following page as peculiarly applicable to what has just been said.

"That the diseases of sheep are numerous and complex is too well known; yet from their extraordinary fewness on some farms, compared with others of the same nature, and even on the same farms under a different management, I am often tempted to conclude that they are naturally as free of them as the hawk or raven; and were I able to define the various parts of the animal frame, their connection with one another, with the influences of climate and regimen upon each of them, I have no doubt but I should make it appear that the whole of the diseases to which this useful animal is subjected, might be traced to have originated in accidence, proceeding from improper usage, or inattention in their keepers or managers. Soils and seasons have their influences, and that to a degree so extensive, as that they will never be entirely bettered; yet still they aay, in a great measure, be guarded against."*

L   It

---

* Hogg's Shepherd's Guide, p. 5.

The only objections I have heard stated to sheltering sheep are the trouble and expence, and that it makes them tender. To the first I answer, the manure and the saving in the lives of many sheep and lambs is a most abundant compensation; and as to the last, it appears to me that an assertion that labourers, fishermen, and others, whose occupations compel them to endure the most inclement weather, ought never to have a dry cottage, or sleep in a warm bed, is entitled to just as much attention. So far from the practice I recommend making sheep tender, I am persuaded that they acquire health and vigour by it, and that they are better able to endure severe weather, when it is necessary, than those always exposed, which generally become ragged in their coats, by tearing off their wool in creeping into brakes and such other places for a warm lodging, when they are not confined.

This admits of illustration from the human subject; for it is a notorious fact, that labourers who live in counties where fuel is scarce, and who consequently are much exposed both to

cold

cold and damp, are much more unhealthy than those who live in districts where coals are plentiful, and who are therefore able to dry their cloathing at night after they return from work.

It appears by some of the statistical accounts of Scotland, that housing sheep is discouraged in that country, though the climate is so much colder; and from the short accounts given of it, as far as I can judge, the discouragement of the very injudicious mode adopted in that country is proper. The sheep appear to have been crammed in close houses, with very little, if any food, and of course must be very unfit to face or bear severe weather when let out, so as to get proper nourishment before the time comes for their confinement again. By such means, the weak state to which they will be reduced must occasion the loss of many lives. The object of housing them appears to be rather making a quantity of manure than any thing else.

The northern farmers adopt a mode of smearing or daubing the sheep with a composition of

grease,

grease, tar, &c. as a means of preserving their health and protecting them from the severity of the climate. Mr. Robert Bakewell, who has recently published a very excellent treatise on the subject of wool, observes, that "by the application of a well-chosen unguent, wool may be defended from the action of the soil and elements, and improved more than can be effected by any other means, except an entire change of breed. Not only the quality of wool will be ensured by this practice, but it will become finer, and the quantity will be increased. It is also found to preserve the sheep in situations where they would otherwise inevitably perish.*"

Being entirely unacquainted with this practice, and as any farther discussion of such matter will exceed my plan, and as it appears to me that Merino sheep, abounding in this greasy, oily substance, require no addition, I shall only observe, that the possession of this quality in some degree accounts for the fineness

of

---

* Bakewell's Observations on Wool, p. 32.

of the wool and the hardiness of the breed. I beg leave to refer my readers to Mr. Bakewell's book, which appears to me to abound with very useful information.

With respect to shearing, my practice has always been to wash the wool upon the sheep's back previous to shearing, and I am persuaded farmers in general adopting this breed upon a large scale will find it by far the best method; for though I have not a doubt of the truth and propriety of every syllable urged by Dr. Parry, and other sensible persons with respect to shearing the sheep unwashed, and scouring the wool in the Spanish method afterwards, (which, notwithstanding what I have to say upon the subject, will be still open to every person inclined to try it,) yet so many things are requisite to be attended to, and so many more hands are required, and so much cookery quite out of the way of the practical farmer, that this circumstance alone would, if it was necessary, deter, I am sure, many from adopting the breed; and in this opinion I conclude I am not singular, as out of the numerous flocks I am acquainted with,

with, one or two gentlemen only, who keep but few, are at that trouble, which, according to Mr. Sheppard, it is hardly worth while to be, and indeed it seems quite unnecessary : agreeing therefore entirely in opinion with him, I prefer making use of his words as better authority. He always washes his flock, consisting in 1806 of near 1000, on the sheep's back, in the common mode practised in this country—and says, " Although the wool of the real Spaniard is so close and compacted together as to admit of but little impression on the grease at the root of the fibre from common washing, yet the dirtier part of the fleece near the surface is considerably cleansed, and the more yolky and pure grease yields easily to the usual process of the manufacturer. In proportion as the cross from the English approaches the Spanish breed, it acquires the same property of yolk; but in every instance that I have seen, it parts much more easily with its grease in the washing. The process of shearing is also much facilitated by the wool having been washed on the sheep's back, which is otherwise very tedious and difficult. To attempt cleaning the

wool

wool after it is shorn, as practised in Spain, would be attended with insuperable difficulties to the grower, were no other objections attached to it; and if left in its full state of grease, it would be very disadvantageous to the manufacturer, as the process of scowering, as practised with the Spanish wool, would be much injured and impeded by the frequent soiling of the liquor used in the operation. It is farther recommendatory of the practice of washing the sheep, that such is in use in Spain with a view to the health of the animal, though not as preparatory to the shearing. I consider also the wool produced in this state, that is, washed from the sheep's back, as in the most merchantable state: it is sufficiently free from excessive grease to enable the manufacturer to judge of its probable waste, which experience will soon render him competent to do; and thus remove a temporary impediment to the sale, which the novelty of the condition of the wool occasioned. The attempt to produce the wool scowered clean would be much more objectionable, as from the inexperience of the party, it would most probably be injured in its softness and quality,'’

quality."* He also says that Spanish wool coming in cleaner condition, adds 6*d*. per pound to its value, compared with that of our own growth, washed on the sheep's back, of equal fineness. Now, if we consider the loss in weight by scowering, according to this statement, nothing of consequence is gained by the practice.

Mr. Bakewell, who is a great friend to anointing *all* sheep, to secure their health and improve their wool, strongly recommends rubbing even Spanish sheep, whose large supply of yolk he admits may make ointment unnecessary, with olive oil on the back and sides immediately after shearing, or with a mixture of olive oil, lard, and wax, to preserve them from cold and wet, and which he says would improve and preserve the soundness of the wool. I have never tried any other method myself than housing, or rather sheltering them for a few nights, if wet immediately succeeds the shearing,

* Communications to the Board of Agriculture, vol. 6, p. 67.

ing, and I have never lost any. My shepherd is very fond of putting the sheep soon after shearing upon fallows or roads, which [he says, from the soil adhering to the skin, in consequence of its oily nature, is almost as good as clothing them; and I think there is good sense in his plan, for turned out as soon as shorn, and continued in green pastures, their skins continue to' be daily washed and cleaned by the dews, which keeps them bare and exposed.

In respect to the effects of climate upon wool, I beg to refer my readers to other authors, and particularly to Mr. Bakewell's book, before mentioned, to which Lord Somerville has appended some judicious notes.

On the effect of food upon wool, as an opinion has long prevailed that fine wool is only to be procured on downs and other short, sweet pastures; and that inclosures, and artificial grasses, as they are called, turnips, cabbages, and all grosser sorts of food, are sure to produce coarse wool, I shall confine myself to the few following observations of persons who have

much

much considered this disputed point, which, I think, set the matter quite clear, and prove the error of such opinions.

Dr. Parry, in his " Facts and Observations,"* first quotes the sentiments of Lord Sheffield, that, " the sheep kept on commons are in general of a bad breed, the wool of which, though commonly reputed fine, is of a much inferior quality to what it is supposed to be ; and it is well known that the wool of ill-fed and neglected sheep becomes of a hairy thread, and that the quantity produced is comparatively inconsiderable. Wool grown on a common is always sold lower than pasture wool of an equal quality." What we are told by the Rev. Arthur Young, in his Agricultural Tour through Sussex, in 1793, (continues the Doctor) is still more strongly in point, viz. " Mr. Ellman's flock of sheep (South-Down) is unquestionably the first in the county. There is nothing that can be compared with it ; the wool the finest, and the carcase the best proportioned ; although

* Page 15.

though I saw several of the noblest flocks afterwards, which I examined with a great degree
of attention; some few had very fine wool,
which might be equal to his, but then their
carcase was ill-shaped; and many had a good
carcase with coarse wool." Speaking afterwards of the food, he says, " Mr. Ellman feeds
his flock with turnips, and has besides other
fine pasture ground; and moreover his sheep
are constantly in the highest order, yet the
wool is remarkably fine." Dr. Parry adds,
" That poorness, or good condition, are not
alone sufficient to produce the difference between coarse and fine wool, we are told, with
regard to the Spanish sheep, by D'Asso, himself a Spaniard, who observes, that the Trashumantes, or travelling fine-woolled sheep, exceed the Estantes, or stationary coarse-woolled
ones, in fatness."

It appears to me, from all which I have been
able to collect, that in order to secure the
finest and best wool, the sheep should be uniformly kept in good store order, which they
ought to be to answer every other purpose;
and

and " in those which are fattened, the increase of weight will make full amends for some dimiuution of fineness. For (says Dr. Parry, in a Communication to the Bath Society) this coarseness is still *only comparative;* and I will engage to produce the fleece of a Merino-Ryeland fat wether, which, though inferior to that of a large proportion of my flock, shall still be finer than the Refina of some of the native Spanish piles."*

The author just quoted, remarks: " There is one point which seems to have been the stumbling-block of many persons who have thought, and of some who have written, on this subject. They observed a sort of gross connection between the food and the quality of the fleece. On one hand, the staple of a sheep which was starved, was weak, and the wool dry, and unprofitable in the manufacture. On the other hand, the wool of sheep on deep inclosed pasture, or on artificial food, was found

to

* Bath Papers, *Vol.* xi. p. 214.

to be coarser and more intractable than that from the Downs. On these two simple facts they thought themselves qualified to reason; and, as is unavoidable from insufficient premises, they reasoned falsely. They concluded that the fine herbage of the Downs necessarily produced fine wool, and that none but coarse wool could spring from gross luxuriant food. Neither of these conclusions is precisely true. The fineness of a sheep's fleece of a given breed, is, within certain limits, inversely as its fatness; and perhaps also (though I am not certain of this point) as the quickness with which it grows fat. A sheep which is fat has usually comparatively coarse wool; and one which is lean, either from want of food, or disease, has the finest wool; and the very same sheep may, at different times, according to these circumstances, have fleeces of all the intermediate qualities from extreme fineness to comparative coarseness. This, which I can demonstrate to be true in the pure Merino, as well as in our native or mixed races, is a principle which admits of but few exceptions. Now if sheep feeding on cabbages, turnips, or oil cake, are only

one

one quarter satiated, and are obliged to move
about so as to dissipate the food which they
have eaten, they will grow thin, and therefore
will have fine wool. On the other hand, if
grass of the finest fibre on a Down shall be
suffered to increase, so that a sheep shall have a
physical power of satiating himself quickly,
without the necessity or the inclination to exercise himself, he will grow fat, and his fleece
will be, *cæteris paribus*, just as coarse as it would
have been on another sheep, which had acquired in the same time the same degree of fatness
from turnips, oil-cake, or any other species of
what is called the grossest food. As, however, a
sheep has less power of growing fat on a down
than in a meadow or turnip field, it is probable
that his fleece will be finer on the former than
on the latter. This is the true cause of that error, which has occupied the mind of almost
every individual in this island, and I might add,
in all Europe, that a Spanish sheep, and probably any sheep out of Spain, cannot yield a
fine fleece. I say in all Europe, because I am
persuaded, that, without this prepossession,
the Spaniards themselves would never have permitted

mitted a single Merino to have been transported from their own country. I conclude from unimpeachable authority, that the contrary is true with regard to the pure Merino race, which, without any deterioration of fleece, lives, in different countries, on every imaginable species of food, not excepting even red herrings; and I know it to be true with respect to the Merino-Ryeland, which, for more than 12 years, I have indiscriminately fed on downs, meadows, hay, ray-grass, clover, vetches, succory, peas, grains, potatoes, turnips, rape, cabbages, linseed, and oil-cake, without being able to discover that the nature of the food has made any difference in the quality of the wool, except through the means of the greater or less degree of obesity which it has happened to produce. I have, however, found that a moderate degree of flesh, such as that of ewes in good store order, which is the state in which I have kept my own flock, will yield wool of that degree of fineness combined with sufficient strength, which makes it best for the purposes of the manufacture of superfine cloth."*

Mr.

---

* Bath papers, *vol.* xi, *p.* 210, *et seq.*

Mr. Bakewell of Dishley, the famous Leicestershire breeder, appears to have been entirely of this opinion, and is said to have made use of these words the year before his death " that he had no doubt that fine wools might be grown on rich pasture lands, by (what he called) overstocking them, and preventing sheep from obtaining more nourishment than they had been accustomed to,"—and the author who mentions this anecdote of him, asserts, " that in proportion to the regularity of the temperature in which sheep are kept, and to the regular supply of nourishment they receive, will the hair or fibre of the wool preserve a regular even degree of fineness."* I beg leave to repeat on this entirely depends their prosperity and the success and profit of the farmer and breeder.

---

* Bakewell on wools, pp. 84, 87.

CHAP.

CHAP. V.

Containing some observations on the disorders to which Merino and Anglo-Merino sheep, in common with other breeds, are liable, though in a less degree, with hints as to the causes and the most approved remedies. Also an improved method of bleeding sheep, &c.

THAT all domestic animals require care and attention, and are abundantly grateful for all they receive, by the profit they afford their possessors, will, I believe, not be denied, and I am of opinion that there is no animal to whom the application of the old adage, " that the master's eye makes the steed fat," applies more properly than to sheep. My flock is managed by a very careful intelligent person; but, on following him over the farm, I have frequently discovered something which had escaped his observation. Sheep, in general, but particularly in hot, showery, or moist weather, cannot be

M   seen

seen too often, as prompt assistance frequently prevents much pain and inconvenience, and sometimes mortality.

It is extremely to be regretted that the health of an animal so material to the prosperity of nations, and particularly of this nation, should hitherto have been so little attended to, and that so few proper remedies are ready to be offered in cases of sickness. The nature and origin of the most fatal, and indeed most of their disorders, are so little known, that when a sheep is taken ill, unless a slit in his ear, a slice off his tail, or opening what shepherds call the eye vein, will save him, he is generally left to perish. My shepherd goes farther than almost any I know, in cases of illness; he gives them a strong dose of salt and water, and they often recover. The above-mentioned practices are probably very often right, but I conclude as frequently wrong. The legislature has given its sanction to the encouragement of the study of the anatomy and diseases of the horse, with much success; and it is to be hoped that some persons of consequence will take sheep also under their patron-

age,

age, and either add them to the veterinary institution, or form another for the express purpose; though the former appears to me, if it can
be effected, the best plan, as such an occupation
would help to fill up the time of those veterinary
gentlemen living in neighbourhoods where attention to horses may not fully employ them. I
have little medical knowledge myself; and shall
therefore only make a few general observations
in particular cases which have fallen under my
notice, or which I have collected from different
authors, and which may possibly not all have
occurred to the practical farmer. I can with
truth add, that no sheep-owner alive, has less personal reason to trouble himself about the matter
than I have, for since I have kept Merino and
Anglo-Merino sheep, I have met with very few
losses. I did not lose one ewe out of 253,
in the yeaning time, and I have as many lambs
as ewes. This I attribute entirely to the natural
health of the race, with the aid of proper shelter
and attention. I have lost one ewe since that
time, whose death was hastened if not occasioned
by being over-driven in treading in oats, sown on
some very light land, as she was before in a weak

state,

state, from having cast her lamb. My wether-hog flock, consisting of more than 100, as I have mentioned in another place, were from necessity extremely exposed, upon some very poor lands on the cliffs near the sea shore. They had the occasional shelter of a hay-rick, while the snow was on the ground only, and of these but one failed, which however is still alive. Of 90 ewe-hogs, or tegs, I lost but one.

Whenever I perceive a sheep getting poorer than its fellows, or attacked by any sort of disorder, whether contagious or not, I instantly remove him from the flock, and put it into good keep and a quiet place; for if left with them, the healthy and strong sheep in general stain and take all the best of the grass or food, and being forced to subsist on what they leave, and consequently to be worst fed, when he ought to be the best, the chance of recovery is much diminished.

The diseases to which the Merino sheep are subject, in this country, according to Dr. Parry, are the rot, scab, hydatids in the lungs, inflammation of the chest, giddiness, and foot rot. To

To which I add what shepherds call the red water, and the white water. And with these may not be improperly mentioned here, the vermin by which they are endangered, viz. the maggot fly, the common little black fly, and the hippobosca, or sheep-tick. Mr. Hogg, the Ettrick shepherd, mentions several other disorders of which I never before even heard the names, and which I fancy are peculiar to some parts of Scotland. In respect to the diseases mentioned above, I do not find any sheep I have yet had, subject to hydatids in the lungs, or inflammation of the chest. I believe it is admitted that some land produces diseases, from which other soils are entirely free.

A variety of opinions are afloat with respect to the rot and its causes, and also the proper remedies, none of which, I am well informed, are to be relied upon. As I am happy to say I know very little of this dreadful disorder from experience, and so many persons have written very copiously on the subject, I shall content myself with quoting Mr. Hogg's sentiments, which appear to me to be rather new, and certainly entitled

titled to consideration. He states, that amongst
the number of opinions which prevail with respect to the real cause of this disorder, he joins
with some of the most sensible and experienced
men in holding it " as an incontrovertible fact,
that *a sudden fall in condition* is *the sole cause of
the rot.*" He says he has been told, that the rot
is occasioned by the sheep living on too soft and
" tothy food, such as grows in wells and
awald lands, or such as are sandy, and have
been fleeted with water;" and that "in one case
this may lead to the cause of it; for the flesh
which the animal acquires by this soft feeding,
not being so firm and permanent as that acquired
by more astringent herbage, such sheep as feed on
the former are much more easily subjected to a
swift decay on the occurrence of any straight;
and this accounts for the rot being most peculiar to soft and grassy soils. But the truth is,
that such lands, instead of being farther the immediate cause of the rot, it is the disease which
induces the sheep to settle upon these. It is no
wonder, says one of my correspondents, that
many people apprehend such food to be instru-
mental in raising the rot; for no sooner is their
constitution

constitution broken by it, than their palate becomes so vitiated, that they delight in nothing else but such garbage as grows about dunghills, cottage yard dykes, and water-fleeted meadows, and this long before their bad state of health is discernible by a great number of people."

"Some say a course of changeable weather, from one extreme to another, raises the rot amongst sheep, and repeat the old proverb— 'Many a frost and many a thaw, soon makes many a rotten ewe.'—This is very true, for there is nothing in the world contributes more to waste sheep than a course of such weather, nor is there any thing more difficult to guard against.

" Others say, that soft weather and a late growth of grass in autumn occasions it. Now, as this is the most tender and soft of all grasses, the former observation is applicable here, that the fat acquired by such feeding is easily exhausted: but this is not all, for it is well known, that a late growth of grass, occasioned by soft weather

at

at the hinder part of the harvest, is ever succeeded by sharp and severe frosts, which wastes this newly-acquired substance with such rapidity, as to gender the seeds of the distemper. Others again say there are two kinds of rot, the black rot and the hunger rot; the one occasioned by foul food, and the other by getting *much too little* food of any kind." He then says, that many gentlemen (whose names he mentions, all able and extensive farmers) firmly assert and prove by many instances, that if you give sheep *always plenty of food,* and GOOD SHELTER, they will never rot; or at least it will never prove destructive.*

In this opinion I very much incline to agree with Mr. Hogg and his friends, from this circumstance: I live surrounded by very poor commons, in many parts full of bogs and wet places, and producing, except a little grass in the bottoms, which are oftener wet than dry, only heath and furze. Of these commons I have made

---

* Hogg's Shepherd's Guide, p. 129, &c.

made a very free use, to relieve my pastures, and prevent their being so much tainted as they would become, from the sheep lying entirely upon them. The sheep, on returning to them from the commons, have eat heartily what they would not look at when they were put out; and though my neighbours who keep sheep, have not ventured to put one upon these commons, and have cautioned me against the practice, I have never lost one, or, that I know of, injured one by it. If I had left them, even for a few days, to live entirely upon the commons, without returning to a belly-full of good food, I have no doubt but I should have rotted all that were so treated.

Happening to call this spring upon a very respectable farmer near Winchester, who keeps a large flock of South-Downs, and enquiring if he was not afraid of rotting his sheep in some water meadows in which I saw them feeding, he informed me that couples (that is, ewes and lambs) never rot in the water meadows, but that he had no doubt dry sheep would; for though he had never rotted an ewe or lamb upon them, his predecessor

decessor had lost numbers of wethers and dry sheep. I have since been told this is a well known fact, though I have not been able to meet with any one who could account for it.

Mr. Hogg remarks that Mr. Price, a very sensible and judicious reasoner, after combating the theory, that moisture is the occasion of the rot, proceeds thus: " The numerous inhabitants of the earth, and sea, and air, are strongly influenced by the seasons and the state of the atmosphere; and the same causes, perhaps, that rapidly call myriads of one species into being, may frequently prove the destruction of another. Is it then improbable that some insect finds its food, and lays its eggs, on the tender succulent grass found on particular soils, which it most delights in? or, that this insect should, after a redundancy of moisture, by an instinctive impulse, quit its dank and dreary habitation, and its fecundity be greatly increased in such seasons, in conjunction with the prolific warmth of the sun? The eggs deposited on the tender grass are conveyed with the food into the stomach and intestines of the animals, whence they are received

ceived into the lacteal vessels, carried off in the chyle, and pass into the blood; nor do they meet with any obstruction till they arrive at the capillary vessels of the liver. Here, as the blood filtrates through the extreme branches answering to those of the vena porta in the human body, the secerning vessels are too minute to admit the impregnated ova, which adhere to the membrane, and produce those animalculæ that feed upon the liver and destroy the sheep. They much resemble the flat fish called plaice; are sometimes as large as a silver two-pence, and are found both on the liver and in the pipe which conveys the blood from the liver to the heart."*

To this Mr. Hogg replies: "That the fluke-worms are found on the livers of all rotten sheep, is a fact, and often in great numbers; but as there has never been any insect discovered on the grass which bore the least resemblance to them, I do not see why we must suppose them taken in with the food, more than that all the worms which breed in the human body are imbibed

---

* Hogg's Shepherd's Guide, p. 132.

bibed in the same manner. Again, as there are no animals subject to the rot but such as chew the cud, it is scarcely supposable that the eggs of an insect can escape into the second stomach so unimpaired as to be capable of being there hatched; for after the food is fermented in the first stomach, upon farther mastication it is so completely bruised and diluted, as to be rendered quite liquid." After some other observations, in which he thinks the foregoing opinion contrary to reason, he says, " It appears to me, that whatever at first produces the fluke-worms on the liver, these are the cause of this particular species of the disease; for, infesting the liver in such numbers, the disease is soon carried from thence to every part of the body in the tainted blood. Now, as salt, or sea-marsh, is well known to prevent and sometimes to cure the rot, this gentleman rationally concludes, that as salt is destructive to all insects, a solution of it given to sheep, when first attacked by the disease, for some time, would cleanse their liver, and quite cure them. Of this he mentions some instances, and in particular, one of a farmer, who cured a whole flock of the rot by giving each

sheep

sheep a handful of salt for five or six mornings successively."*

I also think salt extremely useful in these cases from the following fact.—In February last, I purchased a score of wether hogs, of the second Merino cross, of a gentleman, whose shepherd had so much starved them, I expected many would die, but they all recovered by good nursing, except one, which in the month of April, scoured, and was extremely thin, with a large poke under its chin, and, in short, had every appearance of being in the last stage of the rot. As soon as I perceived this, I took him from the flock, and gave him a strong dose of salt and water, hay, and the best pasture I had: he soon ceased eating the hay: a second dose of salt and water, made as strong as brine, set him up; and he is now returned to the flock, and appears to be thriving.

Another person, mentioned by Mr. Hogg, makes

---

* Hogg's Shepherd's Guide, p. 132. et seq.

makes the following rational observation on this disorder:—" As for the rot, (he says) I maintain that it is always occasioned by too quick a transition from fatness to leanness; and though this discovery may be supposed new, it is nevertheless perfectly correct. There never were any sheep known to rot while they continued *at good equal maintenance*, unless otherwise abused; and none will ever rot on pasture which does not feed them very fat, nor allow them to fall away below a medium. Now this disease can rather be prevented than cured; for this sudden transition towards decay, so completely disorders their whole frame, that to restore it is next to impossible. The substance of the body not having time to be carried off by perspiration, the blood mixes with water, which distils from the flesh, when the consumption commences. This water falls into the veins, and also into the stomach and bowels, and below the tongue. Thus the vitals of the animal are ruined before its body can pine to leanness in a gradual and natural way; while, if its food had been diminished by degrees, and its fat wasted gradually, it might have descended

scended to perfect poverty without any symptoms of the rot being attached to it."*

Mr. Hogg describes many symptoms by which a rotten sheep may be known, but that of the eye appears, on handling them, to be the most infallible one. He relates the following curious anecdote:—A friend of his, on being asked how a man might best judge of sheep by *looking* at them in the fields, where no opportunity offered of examining the eyes, replied to him, " The late Advocate Mackintosh's method of discerning a good man, is the best in the world, whereby to distinguish a sound sheep, his maxim was, ' I never like a man if I do not like his face;' so say I of a sheep, for, if once you take a narrow view of them, *the state of their body is so visibly pourtrayed in every feature*, that you can be at no loss to distinguish them. Their eyes are large and heavy, with a great blotch of white above the star: the top of each ear descends to, at least, a level with the root thereof, and they have each such a
grievous

---

* Hogg's Shepherd's Guide, p. 144.

grievous countenance that no living creature's can equal it."*

The best preventatives of this disorder, according to these authorities, appear to be, what is termed *regular maintenance*, that is, not to suffer sheep to get suddenly poor, or be foul or ill-fed; to provide them with shelter and dry lying; to drain all wet and springy fields, and not to over-stock farms.

A translation of M. Daubenton's Memoir on the diseases of sheep, is given in the work just quoted. His description of their signs of health is very worthy of notice.—It is as follows: " A sheep is in good health when he carries his head high, when the eye is of a clear azure, quick and open eye-strings, and gums ruddy, teeth fast, the face and muzzle dry, the nostrils damp without being mucous, the breath free from any bad smell, feet cool, dung substantial, the mouth clean and of a lively red, all the limbs nimble, the wool firmly adhering
to

* Hogg's Shepherd's Guide, p. 139.

to the skin, which ought to be red, (especially on the brisket) soft, and supple, a good appetite, the flesh reddish, and particularly with good veins, and the hams strong. To know the two latter perfections, the shepherd places the sheep between his legs, and grasps the head with his two hands: with the thumb of the right hand he raises the eye-lid from above the eye, and with the thumb of his left hand pulls down the under eye-lid: he then looks at the veins of the white of the eye; if they are very apparent, if he finds them of a lively red, if the flesh at the corner of the eye, and at the side of the nose, is also of a lively red hue, it is a sign that the animal is in good health. To know if the ham is good, the sheep must be seized by one of his hind legs; if he struggles much to get back the leg, and if much force is necessary to hold it, it is a proof that the animal is strong and vigorous in that part. Sheep are often seen in the market with nose and eyes running, or, as we should say of a horse, almost glandered. This happens in consequence of wet layers during their travel in cold wintry seasons: a continuance of such weather, with

N perhaps

perhaps subsequent neglects, contribute to lay the foundations of diseases, of which, afterwards, the cause is not suspected."

The gid, 'or giddiness, which the French call la tournie, le lourd, is also called the dunt, and, by Mr. Hogg, the hydrocephalus or sturdy. Dr. Parry calls it a rare disease in these sheep, and says he is unacquainted with any cure. Mr. Hogg gives a long and very particular account of it, and proposes a remedy very extraordinary; but, as from the experiments I have myself made, I think it much entitled to attention, I shall fully state his description and ideas of it. " A sheep affected by this disease becomes stupid, its eyes stare, and fix upon some different object from that which it is in fear of; it soon ceases from all intercourse with the rest of the flock, and is seen frequently turning round, or traversing a circle. It is universally allowed that it is occasioned by sheep being exposed too much to rough and boisterous weather, without any shelter."*

The

---

* Hogg's Shepherd's Guide, p. 54.

The immediate cause of this disease is found
to be water contained in a cyst or bag pressing
upon the brain, and according to the Scotch
shepherd, "a bratted sheep (one whose back is
covered with a piece of cloth) will not take it;
and, of a well sheltered flock, very few ever will."
He says the water settles sometimes in one
corner of the skull, and sometimes in another;
and, whenever it begins, it continues to increase
till it is extracted, or the animal dies, when the
brain is generally half wasted. Sometimes it
concentrates in the very middle of the brain,
when it is very difficult to cure; and sometimes
in the hinder parts, where it joins the spinal
marrow, when it is quite incurable. In whatever part of the brain the water settles, the
skull immediately above it becomes quite soft,
and is easily found out. If it is any where in
the crown, the gentlest way is to tap it in the
place where the skull is soft, and let the water
run out, which is commonly performed with an
awl or large corking pin. Dr. Duncan, jun. of
Edinburgh, has invented a small silver trocar,
for the purpose of draining off the fluid, in what-

N 2   ever

ever part of the skull situated, and Mr. Hogg says he has little doubt of its final success. Another old shepherd, who always operated with a large corking pin, has assured him that in thirty years experience, he did not lose one sheep out of twenty on his master's farm, while it was very rare that he could cure any on some of the neighbouring farms. If the skull feel soft in the forehead, then the operation must be performed by thrusting a stiff sharpened wire up each nostril, until it stop against the upper part of the skull. Mr. Hogg remarks, that if this cure were not well authenticated by daily observation, it might seem a very severe and dangerous operation, as the wire goes quite through the brain in two different places; yet a far greater number are cured by this way than any other. The operator must feel for the part of the skull that is soft, and lay his thumb flat and firm thereon, then take the wire, and push it up that nostril that points most direct to the place which is soft, where the disease is seated; and if he feel the point of the wire below his thumb, he may

be

be assured the bag is perforated, but if not, he must try the other nostril. Mr. Hogg thinks this the most certain cure, because the bag being perforated on the lower side, the liquid, as long as any remains, continues to drop through the hole, which it keeps open. He says, he has cured numbers both ways, and killed some in the operation, but most with the wire, which he says arises in some degree from having used the wire when the other means have failed. He has found that leaving the sac, which held the water in the skull, is of no importance, and mentions also several other curious particulars as to this disorder and operation, for which I must refer my readers to his book.*

Of trepanning for this disorder, I am, perhaps, rather more qualified to speak than the last-mentioned writer. Having, this spring, a very valuable pure Merino hog ram, afflicted with it, and being assured after long trial that

---

* Vide Shepherd's Guide, p. 55. et seq.

that nothing but trepanning would save him, and having heard of such an operation being practised abroad with success, I resolved to try it, but deferred it so long, that the animal was at the point of death, having ceased to eat, and being unable to stand; therefore the experiment was made with very little hope of success, but I was anxious to have the operation performed, that I might ascertain the nature of the disorder, having heard it was a grub, which had got into the head, that occasioned it. Indeed, a respectable butcher tells me, they frequently find large grubs in the heads of sheep in the spring, which it is said arise from an insect, or fly, which lays its eggs in the noses of the sheep, and which, when they become larvæ, creep up the nostrils, and lie in the cavities of the head till the spring, when they fall out on becoming full grown. To keep out this fly, is the cause assigned why sheep in very hot weather run with their noses close to the ground, and huddle together. I prevailed on Mr. Charles Palmer, a young gentleman, a pupil of Arthur Quartley, Esq. an eminent surgeon,

surgeon at Christchurch, to perform the operation for me, which he did with such success, that the ram is now as well as any of the rest of my sheep, and in very good condition. The experiment was tried on the first day of March last: in three or four days after it, the ram eat very well; the wound discharged very freely for about three weeks, and for about fourteen days after the usual coloured pus. This sheep was not extremely reduced in flesh, though he had been long ill, at least heavy and stupid, and appeared, by carrying his head on one side, to have lost the sight of one eye. He eat heartily till within about a fortnight of the time of his being trepanned, when my shepherd found him lying down, and unable to rise, and as he thought dying. I took a considerable quantity of blood from him, which, though it apparently afforded him no benefit at the time, was, I doubt not, of great use in preventing inflammation, when the operation was performed. Mr. Palmer, it being his first attempt, made several holes in the skull, two in the front, and one on the crown, before he

could

could find the seat of the disorder, and I understand there was a considerable effusion of blood
at that time. The two holes in the forehead
discharged but little, and were soon well. After the operation, the ram was left without
the smallest hope of recovery, from his being
previously in a state apparently so desperate;
but to comply with my wishes, every thing
was done with the utmost exactness, except
putting a hurdle before the shed he was left
under, which, as he had not stood upon his legs
for two or three days, was thought useless. In
consequence of this omission, the next morning the ram was lost: he had got up in the
night and walked to some distance, and when
found, started, and looked wildly about him,
though he had been so senseless for weeks before, that he would let any one catch him;
he seemed unable to look up, and was indifferent to every thing. Many gentlemen, who
know the circumstance, have been to see him,
but to the credit of Mr. Palmer, the operation
was so neatly performed, that no person could
find out what had been done, or where the

holes

holes were made, unless they were shewn, though it is hardly two months ago.

About a month since, I procured another sheep, on which the same gentleman performed a similar operation, but not with equal success. Indeed, from the following circumstances, it could not be expected. The sheep was extremely emaciated, and the brain much wasted; in fact, it was supposed to be dying when given to me. It lived, however, till the third day after the operation was performed, but never ate any thing. What is most extraordinary, is, the two days it lived, it stood, walked, rose up, and lay down, readily, though before the experiment was tried, it could neither stand nor get up without assistance. Mr. Palmer took a bladder of water entire, out of this sheep's head, and not supposing it contained more than one, completed his operation; but my shepherd on examining the head of the animal, when dead, discovered another bladder of water upon the brain, on the other side the crown, exactly opposite to that which had been removed. The water appears to have been situated

tuated in the two lateral ventricles, or perhaps they were hydatids of the brain.*

I mention all these circumstances in hopes that some other medical gentlemen will turn their attention to this matter, by which it appears to me they may very much benefit society; and young men, by practice of this sort, may acquire expertness in the use of their instruments at the same time.

It may be observed from the following extract of a letter from a clergyman, a friend of mine, who was present when the last operation was performed (and who had not at that time any more than myself heard of Mr. Hogg's book) that the practice of tapping has not been confined

---

* From a very ingenious paper by Dr. John Hunter, in the 1st. volume of Med. and Chir. Transactions, on hydatids in the human body, and from a plate therein shewing the appearance of the hydatids, which I have seen, it so entirely corresponds with the inside of the bag, or sac, taken from the sheep which died, that I am very much inclined to think the disease proceeds from thence.

fined to Scotland. He says, " It has happened to me, in conversing with some of my neighbours respecting sheep, to meet with two gentlemen who have heard of a method of curing
giddy sheep by the practice of trepanning, or
as it may be termed, of tapping the head of the
diseased animal. One of my friends remembers,
when a boy, to have heard of a man who was
famous for curing them by piercing the skull
with an awl, and, as it may be presumed, thus
suffering the oppressing fluid to escape. Another
remembers also to have heard of a man who was
accustomed to make a similar orifice which
he kept open for a few days, and into which
he introduced a quill, for the express purpose
of drawing off the fluid. It seems to me
highly probable that your success in the first
operation with the ram, arose from the brain of
the animal not having been disturbed further
than was absolutely necessary, in puncturing
the bag, and suffering the fluid to escape.
Whereas, in the subsequent operation which
terminated unsuccessfully, the extraction of the
bag itself might make more disturbance to the
brain

brain than so delicate an organ would admit of with safety to the sheep."

The scab is so well-known a disorder, and so many remedies are prescribed for it, by almost all persons who have written upon sheep, that I shall only observe, I think Merino and Anglo-Merino sheep are generally less susceptible of it than other breeds, from the close texture of their wool. But this circumstance may probably render it difficult to eradicate the disease, when from extreme neglect of a shepherd, it is once suffered to get largely into a flock. Some flocks have been much infested with it. Two individuals only, of mine, have had it. They were rams I had let out; and though with the rest, when discovered, they did not communicate it. One of them came home in a very bad state, but they were both cured without difficulty by the following simple remedy, viz.: "A pound of tobacco, on which a gallon of hot water is poured, and close stopped, in the manner that tea is made. The shepherd scratches all the affected parts of the sheep with his nail, till

till he gets off the scurf: he then pours on the tobacco water, coolled, which he works in with his finger, and by one or two dressings the animal in general recovers. In very bad cases, my shepherd bleeds the sheep: he also adds a pint of spirit of turpentine to a gallon of the tobacco water, and gives some tobacco water inwardly, but not that in which the spirit of turpentine is put. The scab is easily discovered. When you perceive a sheep licking or nabbing its wool, and you find any eruption, if you apply your nail and scratch it, the animal appears delighted, and begins to nab immediately, and will bite your coat or any part of your dress he can lay hold of; or he will move his mouth as if chewing.

The foot rot is as well known and as easily cured as the scab, and almost as much prescribed for. My remedy is paring the feet, and applying the following composition:—Oil of vitriol, half an ounce, and sweet oil, one ounce. These are mixed together, and put on with a small stick. I have seldom known it fail after the first or second dressing. As this disorder,

which

which generally commences between the claws, is, as well as the scab, allowed to be very contagious, it is scarcely necessary to add that those infected should be instantly removed from the flock, and to very dry places, for two reasons; one, that the disorder is supposed to originate in low places, or pastures, full of grass, which hold the dews long; and another, that the moisture will wash off, or diminish the effect of the application.

Of the red water, or the white water, I know little more than that they produce almost instant death; principally I suppose for want of proper remedies being known. I mention them here merely with a hope, that should this book happen to fall into the hands of any skilful persons acquainted with their cure, they will have the goodness to make it known.

The red water is an effusion of a red fluid into the cavity of the belly, and is most probably a bleeding from the internal vessels of the belly. It is probably occasioned by high feeding, producing a large quantity

tity of blood. If under these circumstances a sudden check to the circulation should be occasioned by exposure to night air, the blood will be drawn from the external to the internal vessels of the body, and a bleeding may be the consequence; but the blood will be diluted with the watery parts of this fluid, and so give the appearance of bloody water. It is also very likely that the membrane lining the belly becomes inflamed at the same time.

This disease generally happens in the early part of the year, to sheep living well, and after white frosts. About four years ago I lost two South-Down tegs, and two of the first cross, within a fortnight, on some turnips which were running to seed and in a very succulent state. They were bled, but without effect. When dead, they became instantly putrid, so that it was difficult to skin them, the stench was so great. They were full of a reddish water. In May last I lost a well-bred ram lamb at grass, by the same disorder. When discovered, he looked constantly round to his flanks, and bleated as if in violent pain in his bowels. I bled

him

him and gave him salt, neither of which had any effect upon him. These are all I have lost during the last four years by this disorder. If, as is supposed, it arises from luxuriant pastures, my farm has not hitherto been very likely to be much troubled with it; but formerly, on better land, I have lost many Leicestershire sheep by it. It is a disorder which I believe is known on good lands every where.

During the above period I have lost but one sheep by the white water, the symptoms of which are much the same as the red water. Bleeding and a dose of salt and water will sometimes recover the animal in the latter disease; but in the former, bleeding appears rather to hasten the death. When dead with the white water, which appears to be a sort of dropsy, the sheep do not so quickly putrify.

As the only methods of bleeding sheep hitherto practised which I have ever heard of or witnessed, are cutting them on the ear, tail, or under the eye, or on the cheek, as recommended by the Ettrick shepherd, all of which

are

are uncertain as to the quantity taken away, and no mode is ever used, or that I know of, can be adopted to stop the blood, I think it may be useful to mention, that I have lately, with much success, adopted bleeding in the neck, precisely as horses are bled. This was first mentioned to me by my friend the Rev. Mr. Willis, of Sopley, who very obligingly sent his shepherd to teach mine; and it is to be done by any person who can handle a pair of scissars and a lancet, without difficulty or danger. The advantage which it appears to me this mode has over all others is this:—that as all disorders do not require bleeding in the same proportion, nor all subjects, you may by this mode take as much or as little blood as you chuse; and it is easily effected by one person, who, placing the sheep between his legs, with a pair of scissars cuts off a narrow slip of wool in the line of the vein, which he swells by drawing a string placed round the neck, and having opened the vein with a lancet, closes it again with a pin and a bit of wool or horse-hair twisted round it, over the orifice, in the same manner precisely as is practised in horses. Perhaps, as the vein is apt

O   to

to roll, and to elude the lancet, in unskilful hands, a small fleam may be a preferable instrument.

The age of a sheep is known by its teeth. In its second year it puts up two broad teeth in the front of the lower jaw; in its third year it has four broad teeth; rising its fourth year it has six broad teeth; and in its fifth year all its lamb's teeth are cast, and consequently all the teeth are large and broad. These will sometimes remain entire a year or two, and sometimes longer; but it depends chiefly on their food, for after the sixth year they begin to get narrow at the points, wide apart, and are frequently, on Swedish turnips or other hard food, broken; and on heaths and such food, not unfrequently pulled out. Care should be taken in choosing rams to examine that their mouths are even and the teeth proper, and also that the upper jaw does not overhang the lower, which sometimes happens, and makes the sheep bad grazers. Two-toothed sheep, or hogs, or tegs, all of which terms describe a sheep in its first year, require as much attention in the sort of food given them,

as

as old crones or broken-mouthed ewes. The latter, at the time of shedding their teeth (that is, when they are loose and about to fall out) and soon after they are gone, can neither break a turnip nor any hard substance. They ought therefore, to be supplied with grass or good hay. If any appear worse than the rest, their mouths should be examined, and if any irregular teeth are discovered, they should be taken out.

The nature and effect of the maggot fly is too well known to need much description. It is most frequent in pastures much surrounded by groves, high hedges, or plantations; and most active in showery and hot weather, blowing its eggs into, and, as it is called, striking, any part of the wool where it can meet with any moisture, if sheep are not very clean. It generally begins its attacks about the tail and thighs. The eggs become maggots in a few days, or perhaps hours; and, as a moisture exudes from the first nest, the flies continue striking, insomuch, that without attention, the sheep is soon eat to death, for the animal, as soon as they become very painful to it, runs into some hedge, or copse, or hole, and re-

mains

mains there till it dies. A skilful shepherd soon discovers the attack of the fly, by the uneasiness of the sheep, which twists itself about and tries to bite the part. When discovered, the sheep are easily relieved without hurting or cutting off the wool, by squeezing to death, or flirting out the maggots, and putting white lead, or almost any powder, on the place to dry it. This has been my practice hitherto; but I have lately used two new remedies, with very great success, which are described as follows: "Powder for preventing flies from striking ewes and lambs, without injuring the wool, much used by the Warwickshire and Leicestershire breeders, prepared by T. Robinson, chemist, Union-street, Birmingham, and sold in packets of one shilling each, in all the market towns in the county." The other is a liquid, called "Blake and Co.'s Sheep Wash, sold wholesale by Wm. Moore, 82, West Smithfield;" it is also to be had of Mr. Earle, Chemist, Winchester, and in most market towns, and is very effectual.

The above-mentioned powder I find equally useful in keeping off the very small black fly which

which feeds upon any scratch, or sore, or snip
with the shears, or any accident a sheep happens
to meet with. These flies torment a sheep so
much, that it cannot often feed till after sunset;
but none will come near this powder, put on with
a little ointment. The liquid also, put on with a
feather, it is said, has the same effect. Dr.
Parry recommends tar and mutton suet. The
flies, he says, dislike the tar, and the ointment
readily washes out of the wool. Neither the
powder nor liquid at all injure the wool: the
latter destroys the maggots instantly; the powder is only a preventative.

My horned Merino rams have been much infested by maggots at the root of the horn; but
since I have possessed the above-mentioned
powder, of which I put some on all their heads,
as soon as the summer commences, I have not
had one attacked.

Sheep, in different parts of Great Britain, are
certainly subject to many other disorders, some
of which I am acquainted with, such as the
wood-evil, scouring, &c. &c.; but as every
county

county has its remedy, such as it is, and I know of no specific for them all, I shall not extend this work by entering on them, being persuaded, as I have already said, that proper FOOD, SHELTER, and ATTENTION, will prevent more disorders than the most able persons can cure.

FINIS.

Printed by J. M'Creery, Black-Horse Court, London.

AGRICULTURE, &c.

THE Public are respectfully informed, that arrangements are forming for the publication of a New PERIODICAL MISCELLANY, to be exclusively devoted to Agriculture, Rural Affairs, and Political Economy. The primary object will be to disseminate practical information on all subjects connected with Agriculture and Rural Affairs; and on the best means of cultivating and improving our national resources. Further particulars will shortly be announced.

102, Holborn-Hill,
    June 30, 1809.